Beer and Wine Production

A C S S Y M P O S I U M S E R I E S **536**

Beer and Wine Production

Analysis, Characterization, and Technological Advances

Barry H. Gump, EDITOR
California State University, Fresno

David J. Pruett, ASSOCIATE EDITOR
QualPro

Developed from a symposium sponsored
by the Division of Industrial and Engineering Chemistry, Inc.,
at the 203rd National Meeting
of the American Chemical Society,
San Francisco, California,
April 5–10, 1992

American Chemical Society, Washington, DC 1993

Library of Congress Cataloging-in-Publication Data

Beer and wine production: analysis, characterization, and technological
   advances / Barry H. Gump, editor [and] David J. Pruett, associate
   editor.

   p.   cm.—(ACS symposium series, ISSN 0097–6156; 536)

   "Developed from a symposium sponsored by the Division of Industrial
and Engineering Chemistry, Inc., at the 203rd National Meeting of the
American Chemical Society, San Francisco, California, April 5–10,
1992."

   Includes bibliographical references and index.

   ISBN 0–8412–2714–4

   1. Brewing—Congresses. 2. Wine and wine making—Congresses.

   I. Gump, Barry H. II. Pruett, David J. III. Series.

TP570.B45   1993
663'.1—dc20   93–24349
                                                                   CIP

The paper used in this publication meets the minimum requirements of American National
Standard for Information Sciences—Permanence of Paper for Printed Library Materials, ANSI
Z39.48–1984.   ∞

Foreword

THE ACS SYMPOSIUM SERIES was first published in 1974 to provide a mechanism for publishing symposia quickly in book form. The purpose of this series is to publish comprehensive books developed from symposia, which are usually "snapshots in time" of the current research being done on a topic, plus some review material on the topic. For this reason, it is necessary that the papers be published as quickly as possible.

Before a symposium-based book is put under contract, the proposed table of contents is reviewed for appropriateness to the topic and for comprehensiveness of the collection. Some papers are excluded at this point, and others are added to round out the scope of the volume. In addition, a draft of each paper is peer-reviewed prior to final acceptance or rejection. This anonymous review process is supervised by the organizer(s) of the symposium, who become the editor(s) of the book. The authors then revise their papers according to the recommendations of both the reviewers and the editors, prepare camera-ready copy, and submit the final papers to the editors, who check that all necessary revisions have been made.

As a rule, only original research papers and original review papers are included in the volumes. Verbatim reproductions of previously published papers are not accepted.

M. Joan Comstock
Series Editor

Contents

ADVANCES IN HOME BREWING AND WINE MAKING

INDEXES

Preface

BEER AND WINE EXISTED LONG BEFORE humans made their appearance on earth. These beverages were not invented, but discovered, then developed and refined as civilization itself was developed and refined. Wine and beer have been integral components of people's daily food supply since their discovery. In addition, they have played important roles in the development of society, religion, and culture. Indeed, some people mark the beginning of civilization at the beginning of agriculture, and evidence suggests that agriculture began with the cultivation of grapes or grains as much for use in wine making and brewing as for direct consumption as food.

We are currently drinking the best wines and beers ever produced. This fortunate situation is due in large part to our increased analytical knowledge regarding sensory and flavor components, our understanding of the biochemistry and microbiology of fermentation, and the use of modern advanced technology in production. Brewers and vintners produced fine beer and wines for centuries before the biochemical principles of fermentation were discovered. We now know that fermentation is the process by which yeast and natural enzymes convert the carbohydrates in fruits and grains into alcohol and carbon dioxide. The basic biochemistry of fermentation has not changed since prehistory, but the technology used to initiate, control, and guide the process has improved dramatically in modern times. In the past few decades, brewers and vintners have developed and introduced technological innovations that have improved the processes and techniques used in crushing, malting, brewing, fermenting, filtering, packaging, preserving, and stabilizing flavor. Many of these improvements come as a direct result of our rapidly increasing knowledge of the chemical reactions that occur during vinting, brewing, aging, and storing of these beverages.

The traditional brewer and wine maker could prosper with gradual improvements brought about by a combination of folklore, observation, and luck. It was once accepted that 20% of production could be thrown away and made into distilled spirits or vinegar and that a market could be found for even the most marginally drinkable wines. Modern producers face very different economic and market forces. The cost of grapes and brewing materials has escalated dramatically and continues to rise. Consumers, accustomed to large supplies of inexpensive but flavorful and enjoyable wines and beers, demand higher quality products with each passing year. Wine makers and brewers must now rely on science and systematic study of their art to develop the technology and innovations needed to meet the economic challenges they face while pleasing the increasingly knowledgeable customer base that enjoys beer and wine as a part of civilized society.

Chemists, chemical engineers, and specialists in other fields with extensive training in chemical principles and techniques have made significant contributions to the development of our modern understanding and technology. Hence, it was most appropriate to organize a symposium on advances in the analysis, characterization, and technology of wine and beer production in San Francisco, near the heart of one of the world's great wine-producing regions. We have observed that a substantial fraction of our colleagues in chemistry, chemical engineering, and related disciplines are interested and sophisticated consumers of these beverages. The multidisciplinary nature of brewing and vinting means that much of the information that would be of interest to these "technoconsumers" is scattered in the diverse literature of fields such as food science, agriculture, microbiology, and analytical chemistry. Thus, this volume is intended to consolidate some of the most recent information. The primary target audience for this book is chemists and others with both a strong technical background and an interest in wine, beer, or both who have no specific, first-hand knowledge of the quality concerns of the industry or how modern technology is used to produce the high-quality products available today. A secondary audience is the students and professionals in the wine and beer industries who wish to view the status of current technology.

This book addresses a number of aspects of wine and beer production, characterization, and analysis. For some wine and beer drinkers, sensory and physicochemical evaluation of their favorite beverage is no more complex than "it smells and tastes good (or bad)." Indeed, we would be the last people to argue that any more in-depth analysis is required to enjoy a good glass of wine or beer. Even nontechnical wine and beer drinkers have achieved technical sophistication in the past decade. Conversations at beer and wine tastings conducted for the general public are sprinkled liberally with references to pH, total acidity, Bret, and cold stabilization. Certainly the average consumer is not required to understand the science and technology of beer and wine production to enjoy these beverages. However, many consumers find that such an understanding leads to greater appreciation of the products.

Acknowledgments

We thank the Division of Industrial and Engineering Chemistry, Inc., of the American Chemical Society for encouraging and assisting in the organization of the symposium from which this book is derived. One of the high points of the symposium was the excellent and informative wine tasting hosted and directed by Wilford Wong of San Francisco; for his efforts we are most appreciative. Personnel at ACS Books have been most helpful in providing a schedule and the necessary reminders for completion of the various chapters and for editorial and technical assistance in the preparation of the final volume. D. J. Pruett gratefully acknowledges QualPro for its continuing support of this and other professional activities within the ACS. Finally, we are most especially grateful for the cooperation, support, and encouragement of the symposium speakers and authors who produced this book. Without their special efforts and willingness to meet externally imposed timetables, we would not have this book today. We hope that everyone involved in the production of this volume will, with the readers, find many hours of education and enjoyment in these pages, perhaps most appropriately accompanied by a glass of fine beer or wine.

BARRY H. GUMP
Department of Chemistry
School of Natural Sciences
California State University
2555 East San Ramon Avenue
Fresno, CA 93740–0070

DAVID J. PRUETT
QualPro
P.O. Box 51984
Knoxville, TN 37950–1984

May 25, 1993

ADVANCES IN ANALYSIS
AND CHARACTERIZATION TECHNIQUES

Chapter 1

Technological Advances in the Analysis of Wines

Gordon H. Burns[1] and Barry H. Gump[2]

[1]E T S Laboratories, 1204 Church Street, St. Helena, CA 94574
[2]Department of Chemistry, School of Natural Sciences, California State University, 2555 East San Ramon Avenue, Fresno, CA 93740-0070

Primary goals of wine analysis include reducing spoilage, improving yield, efficiency and wine quality and meeting export specifications. Components typically determined in wines are soluble solids, acidity, alcohols present, carbonyl compounds, esters, phenolic compounds, chemical additives, trace metals, oxygen, carbon dioxide, fluoride and certain amino acids and protein. Analytical capabilities in the wine industry and in service laboratories range from the primitive to most sophisticated, and wineries choose their analytical methods based on the precision required and volume of analyses to be performed. In addition to traditional wet chemical methods, laboratories utilize flame AA, gas chromatography and GC/MS. Recent major advances in wine analysis methodology include the use of automated batch analyzers, near-infrared spectroscopy, C13 NMR, and isotope-ratio mass spectroscopy and the development of methods of characterizing wines via their "signatures" or "fingerprints."

Table I presents an overview of the history of wine production and analysis. The beginnings of wine production and the advent of analytical methods applied to winemaking are noted. Some of the high water marks include Pasteur's 1858 discovery of acetic acid forming bacteria and the role they play in wine production. Following Prohibition, the development of wine technology and governmental regulation increased in scope. This meant, coincidentally, an increase in demand for analyses related to regulatory requirements. Following Prohibition it became more important for winemakers to rely upon science in addition to art and chance in the production of their wines. Relegating even part of their production to distilling material or vinegar production became increasingly difficult due to the escalating cost of grapes and production. These factors, too, drove the need for improved analytical techniques in winemaking.

Table II provides an overview of why one might be interested in analyzing wine. The first set of reasons involves quality assurance. Quality assurance analyses are performed during the ripening of grapes and continue through the entire processing cycle. Indeed, such analyses follow the wine after bottling, for as many as 40 to 50 years after reaching the bottled, finished state.

The goals of wine analysis include 1) reducing spoilage, 2) improving the process with regard to yield and efficiency, and, in particular, 3) improving the quality of wine. Every year the number of wines in the marketplace increases significantly

0097–6156/93/0536–0002$06.00/0

TABLE I
History of Wine Production and Analysis

- 3500 B.C. & earlier: Assyrians, Egyptians, Iranians

- 1400 B.C.: Greeks (uin-, oinos)

- Romans & Hebrews: "new wine into new bottles"

- Romans carry viticulture to France, Germany, Spain, etc.

- 400 A.D.: dark ages - Syrah to France, sacramental wine

- 1858: Louis Pasteur - Acetobacter, science & wine

- 1870: Phylloxera (root louse), America's gift to Europe

- Post-Prohibition: quality improves, regulation begins

- Industrialization: technology explodes, odds improve

TABLE II
Overview of Wine Analysis

- Quality Control: ripening, processing & aging

- Chemistry & microbiology complement sensory analysis

- Spoilage reduction, process improvement, highest quality

- Blending: precise analysis = precise blends

- Industry capabilities: refractometer through GC/MS

- Export certification: E.E.C., Pacific Rim, Canada

- Reference methods: A.O.A.C., O.I.V., E.E.C., industry

- Choice of methods: "Why doing it, how many to do?"

- Sampling: grapes, must, production & aging, bottled wine

and the challenge to the winemaker is to increase quality in order to better compete in that marketplace.

It is important to remember that chemical and microbiological analyses only complement sensory analysis. We know in the laboratory and in daily experience that our most sensitive analytical tool still remains the nose. Even as compared to many sophisticated GC/MS analyses, we are still able to detect compounds at lower levels with our noses.

Blending is one example of an area where precise analytical techniques are required. Precise blends require precise analyses. For instance, many wineries use gas chromatographic analyses for ethanol determinations to determine if their tanks have been blended consistently. These analyses typically produce results with a standard deviation of less than 0.02 percent vol. ethanol.

Analytical capabilities within the industry and in laboratories serving the industry vary widely. The federal government requires that all commercial wineries have, at minimum, two pieces of equipment: One device for determining ethanol content of their product and the other for gauging the fill level of their product. The most common device for ethanol determination found in small wineries is the Ebulliometer. This device measures ethanol content by boiling point depression. The very simplest means of determining fill-level is by using an accurate and certified flask graduated to indicate the most common bottle sizes. The most sophisticated wineries and service laboratories carry capabilities up through GC/MS and beyond.

Another principal reason for analytical work with wines involves export certification. Countries of the European Economic Community have a set of requirements for analyses to be performed prior to import as do the Pacific Rim countries such as Japan, Korea, and Thailand. Canada, too, has an extensive list of analytical requirements that are performed either before export to Canada or by the Provincial Liquor Control Boards upon arrival.

Reference methods within the wine industry are those defined by the Association of Official Analytical Chemists (AOAC) and the Office International de la Vigne et du Vin (OIV). The latter organization is a worldwide body of governments of wine producing countries; The European Economic Community adopts OIV regulations as its own. In addition to these official methods, there are commonly accepted reference methods within the wine industry. Some of these are defined by texts from the enology departments at educational institutions such as California State University at Fresno and the University of California at Davis.

The choice of analytical methods for wineries is very much influenced by the precision required and volume of analyses to be performed. A technique that may be appropriate for the largest producer in the country dealing with hundreds of samples per day with a high degree of accuracy may not be appropriate for a small family-owned winery that infrequently runs a few samples. This may seem obvious, but it is a point that is often ignored by winery laboratories when choosing analytical methods. In any discussion of analysis, it must be emphasized that analytical results will be only as valid as the samples from which they are derived. Sampling is both a critical and difficult process in the wine industry. Sampling grapes, as one might imagine, is a prime example. We talk about the concepts of the "100 berry" sample or the "three cluster" or the "five cluster" sample. Many of the decisions that have to do with selecting grapes, determining their suitability for winemaking, and time of harvest are made not only on a sensory basis but also on the basis of analytical results. As the production process continues, sampling perhaps becomes easier but still produces new challenges at each stage. When dealing with must or juice before fermentation, issues such as settling, mixing, and maceration of the pulp must be taken into account. When dealing with wines from multiple containers such as barrels, attention must be paid to mixing within the barrel and appropriate statistical concepts to result in uniform samples from a number of individual containers. With bottled wine, factors such as the storage history of individual bottles and the condition of the closures must be considered.

The components of wine and must can be broken into classes requiring analysis (see Table III). The first of these classes is that of soluble solids. Soluble solids means that which we loosely refer to as "sugar," which is in fact the sum of individual sugars and other non sugar components. One may often read on wine labels that the "sugar" at harvest time was, for example "24.5 degrees Brix". It must be remembered that Brix is in fact not a measure of sugar at all, but rather the refractive index or other measurement of soluble solids, which is often presumed to include only sugar. This is certainly not completely correct, but it generally serves the practical needs of the industry. Extract or total dry extract is a measure of the residual solids left in a wine. It is a determination routinely performed within the industry, especially for import and export purposes. Many countries have formulas within their regulations regarding ratios of extract to ethanol in the finished product. These formulas are designed to detect fraud and "watering" of grape musts or wine during production. The most commonly determined individual sugars in wines are glucose and fructose, the two primary fermentable sugars in wine. A few of the several techniques available for this determination will be mentioned later.

Acidity in wine is expressed in various ways. The most common is total acidity or, more correctly, total titratable acidity. Volatile acidity, which in the case of wines is almost exclusively acetic acid, is determined as a measure of spoilage. Wine pH has important sensory and stability implications and consequently is closely monitored. Changes in wine pH of as little as 0.02 - 0.03 pH units can be significant. Individual acids present in wine are also frequently determined by winemakers. In the berry and juice before fermentation, these acids include tartaric, malic, and traces of citric acid. As the juice ferments and becomes finished wine, the list grows to include tartaric, malic, lactic, acetic, citric, and succinic acids. Traces of other organic acids are present as well.

Among alcohols present in wine, clearly the major component is ethanol. Other alcohols present include small amounts of methanol, glycerol, and various fusel oils. The significance of these fusel oils is often underestimated, since they can in certain cases greatly influence the character of the overall product.

Carbonyl compounds present include acetaldehyde, which generally occurs in elevated levels after oxidation of a wine. Hydroxy methyl furfural is produced from the dehydration of fructose. It is often present in elevated levels in wines that have been baked. It also is often measured in ancillary products of the wine industry such as juices and concentrates. Another carbonyl compound is diacetyl, which is one of the main identified sensory components resulting from a malo-lactic fermentation.

Esters present in wine include ethyl acetate, generally referred to by wine writers as the "airplane glue" smell. Ethyl acetate often results from hot fermentations and uncontrolled microbial activity. Another ester is methyl anthranilate, a component of Concord grapes. Anyone who wishes to identify this compound in a favorite table wine should first have a smell of one of the many Concord Grape juices or Concord grape flavored products.

Nitrogenous compounds include ammonia, one of the principal and most easily assimilated nitrogen sources used during fermentation. There are many and varied amino acids present in wine, the principal ones being proline and arginine. Common thinking has traditionally implicated elevated levels of amines, such as histamine, in the headache problems wines are thought to cause. Much research has been done that should have debunked this theory, but it remains a matter of fact that some countries, such as Switzerland, reject wines that have histamine levels exceeding 10 mg/L. Proteins also exist in wine. Generally for reasons of stability winemakers attempt to remove these proteins during production by fining with agents such as bentonite.

Many of the desirable and unique characteristics of wine can be attributed to phenolic compounds. Those measured routinely in wine include a group described as total phenolics, using a colorimetric method. Also determined on a less frequent basis are individual anthocyanins, (the color components of red wines), and individual phenolics using HPLC.

Chemical additions to wines that demand analysis include sulfur dioxide. Sulfur dioxide has been used as a preservative and has been present in wine throughout the history of wine production, which spans at least 3,000 years.

Furthermore, in certain wine products, but generally not premium wines, preservatives such as sorbic and benzoic acids are used and therefore require analytical methods for their determination. In general these preservatives have been used with very good results and minimal deleterious effects. Scandals do occur occasionally in the production of wine, though generally not involving United States wines. Several incidents have occurred in Europe in recent years. Some of the tragic ones involved the addition of methanol to wines in Italy and the addition of diethylene glycol in Germany and Austria.

Other components of wine that laboratories deal with include common and trace metals. The common metals include calcium, copper, iron, and potassium. In addition to these there has been some focus recently on heavy metals such as lead and cadmium, which certainly exist at trace levels in wine as they do in all other natural food products. It is interesting to note that the levels found in wine are usually closer to levels found in typical drinking water than they are to those in other foodstuffs.

Oxygen is monitored in wine to prevent certain problems that can occur when its level is high. Carbon dioxide is certainly measured in the case of sparkling wines, and specified levels of CO_2 often exist in other table wine products. CO_2 levels are becoming something of a stylistic tool in some wine production today. An official titrimetric method for measuring CO_2 exists but the sampling is very technique-dependent and therefore rather difficult for most wineries to use and obtain results with acceptable accuracy.

Many other compounds are analyzed routinely in wines. One of these is fluoride. Limits have been placed for acceptable fluoride levels by the international community and are currently enforced by Scandinavian countries. Fluoride occurs naturally at trace levels in wines. Principally in the Central Valley of California an organic pesticide called Cryolyte is used. This fluoride containing material originates in Greenland as a mineral mined from the ground, is processed in Norway, and is shipped throughout the world for use in controlling certain leaf devouring insects. In spite of its organic certification and innocuous mode of use, traces of fluoride can sometimes remain in finished wine products when this pesticide is used. Wines containing fluoride at above one mg/L have run afoul of regulations.

A wide range of technology is used in current analytical techniques and applications (Table IV). The traditional analytical methods have been based on wet chemical techniques. If one were to walk into a wine laboratory in 1890 and examine the equipment, it would not appear to be much different from the equipment used in most winery laboratories today. The techniques used now are well-developed and have been tested over the years to produce acceptable and usable results. Many of the modifications currently being made to these proven and time tested wet chemical methods are to translate them into automated versions.

HPLC (High Performance Liquid Chromatography) is used in a number of winery laboratories (to put this in perspective, probably not in more than four or five within the United States) for the analysis of various compounds. Probably the first commonly accepted application of HPLC in the winery laboratory was for the analysis of organic acids. Techniques for this are many and varied. All generally produce acceptable results and involve varying degrees of sample preparation. None of the valid methods to date involve no sample preparation. HPLC techniques also exist for

TABLE III
Wine and Must Components

- Soluble solids: "sugar," extract, glucose & fructose

- Acidity: total, volatile, pH, individual acids

- Alcohols: ethanol, methanol, fusel oils, glycerol

- Carbonyl compounds: acetaldehyde, HMF, diacetyl

- Esters: ethyl acetate, methyl anthranilate (labruscana)

- Nitrogen compounds: NH3, amino acids, amines, proteins

- Phenolic compounds: total, anthocyanins

- Chemical additions: SO2, sorbic & benzoic, scandals

- Other: common & trace metals, oxygen, CO2, fluoride

TABLE IV
Analytical Techniques & Current Applications

- "Wet Chemistry:" manual vs. automated

- HPLC: acids, sugars, phenolics, reality check

- AA: Cu, Fe, Ca, K, trace elements including Pb

- GC: EtOH,MeOH, higher alcohols, esters, DEG

- GC/MS: ethyl carbamate, procimidone, sulfides, 2,4,6-TCA, pesticide residues, contamination

- NIR: EtOH, "residual sugar"

the determination of sugars. Usually, for wine making purposes, glucose and fructose are analyzed to determine the degree of completion of fermentation. However, for certain wine types the analysis of sucrose is performed for regulatory purposes. It should be noted that at wine pH (3 - 4) sucrose is gradually hydrolyzed to glucose and fructose. HPLC is applied mostly on a research basis for the analysis of various phenolic compounds. This is certainly one of the areas where HPLC may prove to be a benefit to the industry. When considering the use of HPLC in regard to wine analysis, it is important to apply a reality check and question why the analysis is being done and how many samples have to be run. For running many different wine analyses with limited numbers of samples for each, as is necessary in a typical winery environment, HPLC is often not the tool of choice. In those situations where there is sufficient sample demand to justify a dedicated instrument with auto sampling capability, HPLC methods have potential merit.

Flame atomic absorption is routinely used for the determination of metals, such as copper, iron, calcium, and potassium in wines. Other trace elements are determined using flame or electro thermal atomization (graphite furnace) techniques. The principal trace metal of high interest today is lead.

Gas chromatography is used routinely by a greater number of wineries than those that use liquid chromatography. The principal application for gas chromatography is that of the determination of ethanol. There exists an AOAC-approved method for the determination of ethanol in wine, and the technique is very precise and accurate. Methanol is determined fairly regularly for regulatory and production purposes. Higher alcohols and fusel oils are determined routinely in wines and also in wine byproducts, such as brandy. Also determined by GC are various esters and certain other compounds, such as diethylene glycol, for regulatory purposes.

Gas chromatography-mass spectroscopy (GC/MS) is used increasingly in the industry (though in only a small number of industry laboratories) for the determination of compounds such as ethyl carbamate, procymidone, and other compounds listed in Table IV. Ethyl carbamate, (urethane), is a compound that has recently been picked up as a tool by various neo-prohibitionist groups to attack not only wine but other alcoholic beverages. Ethyl carbamate is formed whenever urea, (ubiquitous within our environment), and ethanol come into contact. The reaction is accelerated in the presence of heat. Ethyl carbamate consequently exists in extremely trace levels in wines. Although there has been much talk, no scientific research has been done to date that gives any valid evidence of harmful effects of ethyl carbamate on humans. GC/MS is also used for the determination of Procymidone. Procymidone is a fungicide used widely throughout the world on grape crops. Unfortunately, it was not registered by its manufacturer for use on grapes within the United States. Consequently there has been a large demand for procymidone analyses because wines containing any analytical trace of the compound, even at levels 100-fold lower than the European regulatory limit, have been banned from importation into the United States.

Wine laboratories have developed techniques for analysis of sulfides in wines using GC and GC/MS. This is one of the more interesting applications of GC research and is part of an attempt to back up sensory examinations of wines with more specific analytical information.

One of the more interesting compounds that the wine industry has been forced to deal with recently is 2,4,6-trichloroanisol. This is one of many compounds sometimes present in wine corks that can apparently contribute in certain cases to a sensory characteristic described by many as "corkiness." This compound presents a significant analytical challenge since its sensory threshold in a typical wine matrix is around two parts per trillion.

Wine as a product is not immune to the current regulatory concern regarding pesticide residues. Certainly pesticides, fungicides, and various other agricultural chemicals are used on wine grapes in varying amounts. It is interesting to note that, because of the scheme of wine production, it is extremely unlikely that significant levels of pesticide residues would remain in the finished product. Many of these residues decompose and/or precipitate even at the stage prior to the onset of fermentation. Most of the remaining residues are readily assimilated by yeast during fermentation. Consequently they are settled or filtered out with the yeast cells. The very trace amounts of residues remaining after fermentation are often insoluble in a solution containing 10 to 15 percent ethanol, such as wine. Nevertheless these analyses are continuously being requested.

GC/MS is often the tool of choice for detection of contamination during processing. It is important to note here that, as discussed previously, minute contamination that cannot be analytically detected in wine, even with sophisticated GC/MS techniques, might still be detectable by sensory means.

Recent major advances in wine analysis methodology (Table V) include the adaptation of automated batch analyzers from the clinical field for wine analysis. Such analyzers usually utilize colorimetric or enzymatic techniques and provide rapid and extremely precise analyses of certain wine parameters with a high degree of high quality assurance integrated. Parameters that have been adapted to such methods include glucose and fructose, various organic acids including tartaric, malic, lactic, acetic, and citric, total phenolics using various colorimetric techniques, urea, ammonia, and other compounds. Clearly, to the extent that such analyzers can be trained to perform the remaining menu of analyses applicable to the industry, the future of wine analysis will flow in this direction. Other current interim approaches are based upon flow techniques such as segmented flow analysis, and flow injection analysis. These techniques currently provide for automation of several analytical procedures, such as on-line pre treatment by dialysis or diffusion. Such procedures are currently not amenable to modern batch analysis techniques. Laboratory experience has demonstrated reasonably successful use of flow injection and segmented flow techniques.

An increasingly popular technique is the use of near infrared spectroscopy (NIR). Instruments are commercially available allowing the determination of ethanol in wine at equal or even better accuracy than that attained using GC methodologies. An AOAC collaborative study is currently being initiated for the determination of ethanol in wine using NIR. Future applications of this technique will also include the analysis of residual sugars in wine. Dr. Karl Norris, who analyzed various agricultural products at the USDA laboratory in Maryland during the 1960s, performed some of the earliest analytical work using NIR. By using statistical correlation techniques suitable for measuring the concentration of oil and protein in grain, Dr. Norris and his colleagues identified a group of wavelengths. These specific wavelengths are still used in most NIR analyzers for a wide range of applications, including the measurement of ethanol in wine. One of the major advantages of this technique is that no reagents are required. This is particularly important given hazardous waste disposal requirements for most laboratory operations today. The techniques employed in NIR to obtain infrared measurements of a sample include specular or direct reflection from a sample surface, diffuse reflection, and more recently, transmission. Transmission is the principle on which the currently successful near infrared analyzers operate. Absorption bands used in wine analysis range in wavelength from 1700 to 2300 in. In general, hydroxyl groups for alcohol determination and various types of vibrations including stretch and bend/stretch combinations are examined. In addition to the primary absorption band, the first and second overtone bands are taken into account. As few as three wavelengths can result in acceptable ethanol data in a wine matrix. In practice, six wavelengths are typically measured incorporating the wavelengths necessary for measurements of both ethanol and sugar. Near infrared measurements are extremely dependent on temperature. Consequently, successful instruments incorporate either a constant temperature water bath or, in more modern instruments, an electronic Peltier temperature control that allows for control within a few tenths of a degree Celsius.

Techniques for analysis of trace metals in all matrices including wine have been improved in recent years. These improvements are based on the use of more reliable platform surfaces in electro thermal vaporization chambers and advancements in auto sampling techniques. In providing greatly improved, reproducible sample introduction, these auto sampling methods eliminate many of the problems that can occur with manual sample introduction. Work is also in progress to attempt to improve detection limits currently available with inductively coupled plasma (ICP) techniques. At present these techniques do not allow for the determination of certain important metals with acceptable sensitivity in the wine matrix. Should detection limit targets be met, then the obvious advantages of ICP would be the simultaneous determination of several important metals in wine.

Some future applications and the industry "wish list" can be seen on Table VI. In Europe, principally in France at the University of Nantes, researchers have done extensive work with regard to authentication of wine using C13 NMR and isotope-ratio mass spectroscopy (IRMS). The people working in this field in France report to have been able to differentiate not only sources of wines but even vintage years based on their IRMS and C13 NMR techniques. The OIV is currently discussing authenticity regulations enforceable by these techniques. The United States' wine analytical community has questions concerning the applicability of these techniques, but we will undoubtedly hear more about them in the future.

As was mentioned earlier, it is probable that near infrared techniques for total phenolics, and possibly phenolic fractions, may be attainable. Much work remains to be done with regard to the NIR analysis of fermentable sugars in wine. The biggest challenge is the determination of these compounds in wine at the low concentrations necessary for determination of dryness or completion of fermentation.

Today's challenges for HPLC are to make reality meet the expectations of the wine industry. HPLC is an extremely powerful technique, especially when coupled with auto sampling capabilities, that has been and will continue to be used within the industry. The challenges are to make improvements in this technique, primarily in the areas of sample throughput versus time required for analysis, analytical reproducibility, column life, stability of retention time and selectivity.

Also mentioned earlier was the problem of cork taint in wine. This problem is perceived as so severe by the consumer that some are suggesting wooden corks be replaced entirely as a closure for wine bottles. Whether this will finally occur or not remains to be seen. In the meantime, analytical techniques are required for determining all of the compounds, in addition to 2, 4, 6-trichloroanisol, which may be responsible for cork taint.

Sulfides in wine have always been and continue to be an interesting problem. Wines are occasionally bottled with no apparent sensory problems and only months later develop perplexing and annoying sulfide aromas. These aromas can frequently be attributed to very highly reactive sulfide compounds such as diethyl and dimethyl disulfides. Analytical methods exist for the determination of these compounds at the detection limits required, but the complexity and expense of these methods precludes their use on a routine basis.

Regulatory compliance with the wine industry has traditionally been a relatively insignificant matter. Historically only very few compounds have been regulated in the wine matrix, but this number is now beginning to escalate. The demand for analyses for trace metal content and agricultural chemical residues will surely continue to grow.

TABLE V
Recent Major Advances in Methodology

- "Discrete" automated batch analyzers

- Segmented flow analyzers (CFA, SFA)

- Flow injection analyzers (FIA)

- Infrared analyzers (NIR)

- Improved equipment and techniques for trace metal analysis

TABLE VI
Future Applications & Industry "Wish List"

- Phenolic compounds:
    a practical HPLC approach?

- Cork "taint:"
    practical screening for corks?
    identification of responsible compounds?

- Sulfides:
    can we supplement sensory impressions?

- Wine "signatures:"
    NIR?
    combined techniques?

Over the past ten years attempts have been made to characterize wines using techniques often referred to as wine "signatures" or "fingerprints." Many of the proposed techniques have verged on being fraudulent and yet the need for such techniques is recognized. Uses within the industry for such a fingerprint or means of identification of a particular wine blend or lot are potentially many. An example of this would be wine in trade. Suppliers would often like to tag their wine or identify it with an analytical signature to assure their buyers that the wine they received is the wine they purchased. This "signature" concept could conceivably trace authenticity all the way back to the grape variety including clonal selection, appellation, vintage, etc. Since this is of major interest to the industry, we can expect to see significant developmental efforts in this area in the future.

Conclusion

For the first 3,000 years or so of wine production, analytical characterization of wines was primarily conducted using sensory analysis. Laboratory analytical methodology has traditionally been selective or sensitive enough only to provide information on the major constituents in wine. Today this situation is changing. Quality assurance and regulatory requirements are forcing wineries to provide more analyses for both major and minor components. This trend can only grow as the industry becomes more regulated, especially in regard to agricultural chemical and trace metal residues potentially remaining in a finished wine.

While significant numbers of analytical measurements are still being made using traditional wet chemical methods, these are in the process of being automated in more advanced laboratories. Automation has the advantages of increasing sample throughput and improving relative precision and accuracy of measurement. In the interest of producing a higher quality product there are a number of instances in which instrumentation is available and utilized to provide analytical measurements at the trace levels required for monitoring of sensory characteristics of wines. This trend will continue, as winemakers seek ever more definitive information on the products they produce. New instrumental approaches to analyses required by the wine industry will also continue to expand. The use of near infrared techniques without the need for sample preparation or reaction reagents is an important example of this. Carbon 13 NMR, isotope-ratio mass spectroscopy, GC/MS, and other instrumental techniques will continue to be used to explore the concept of grape and wine "fingerprints" for purposes of authentication.

Bibliography

Amerine, M.A.; Ough, C.S. *Methods for Analysis of Musts and Wines.* John Wiley & Sons: New York, NY, 1988.

Zoecklein, B.W.; Fugelsang, K.C.; Gump, B.H.; Nury, F.S. *Production Wine Analysis.* Van Nostrand Rheinhold: New York, NY, 1990.

Official Methods of Analysis, 15th ed. Association of Official Analytical Chemists: Arlington, VA, 1990.

Recueil des Méthodes Internationales d'Analyse des Vins et des Moûts, 5th ed. Office International de la Vigne et du Vin: Paris, 1990.

RECEIVED May 13, 1993

Chapter 2

Advances in Detection and Identification Methods Applicable to the Brewing Industry

T. M. Dowhanick and I. Russell

Labatt Breweries of Canada, 150 Simcoe Street, London, Ontario N6A 4M3, Canada

Several alternative techniques to 'classical' microbiological detection and /or identification of potential beer spoilage microorganisms are reviewed. In general, the major benefit of using the described techniques is a very significant time reduction (from days or even weeks, to hours or days respectively) in obtaining the assessment. In some cases, portability becomes an additional benefit, while in other cases, automation is a key factor. Not all of the techniques described are either user-friendly, inexpensive or have been optimized for utilization in a Quality Control environment. However, they do offer a glimpse into the not-too-distant future of how microbiological assessment of brewery samples will most likely be conducted.

In the brewing industry, as is generally the case with food and beverage companies, failure to keep potential beer spoilage microorganisms to the minimum practicable level can lead to economic losses as a result of reduced product shelf life and inconsistency in product quality. For a variety of reasons including increased consumer awareness on product quality, tightened government regulations, increased competition among brewers due to declining consumption, the increasing trend to avoid pasteurization of packaged beer, and technological advancements, the last decade or more has witnessed the development of a plethora of novel or improved methods for the detection and/or identification of microorganisms.

For which microorganisms must the brewer monitor? Factors such as the presence of alcohol, sub-physiological pH, anti-microbial hop components, and an anaerobic environment inhibit the growth of human pathogens in a typical 4% to 5.5% ethanol beer. As a result, the scope of microorganisms typically isolated in a brewery remains limited when compared to the multidiverse array of microflora found in the medical, environmental, cosmetic or food and dairy industries. Nonetheless, the detection and identification of brewery microorganisms is not always a simple or straightforward procedure. Among the Gram positive bacteria known to establish residence in breweries are the lactic acid bacteria (*Lactobacillus, Pediococcus, Leuconostoc,* and *Streptococcus* spp.), along with members of the

0097–6156/93/0536–0013$06.00/0

genera *Micrococcus,* and *Bacillus.* The Gram negative brewery bacteria include the acetic acid bacteria (*Acetobacter* and *Gluconobacter* spp.), the enterobacteria (*Obesumbacterium, Enterobacter, Citrobacter, Klebsiella, Hafnia* and *Proteus* spp.), *Zymomonas, Pectinatus,* and *Megasphaera* as well as *Alcaligenes, Acinetobacter, Flavobacterium* and *Pseudomonas* species. 'Wild' yeasts, comprising either non-brewing or (when isolated at locations not desired) brewing yeasts, include members of the genera *Brettanomyces, Candida, Cryptococcus, Debaryomyces, Endomyces, Hansenula, Kloeckera, Pichia, Rhodotorula, Saccharomyces, Torulopsis* and *Zygosaccharomyces* species *(1-7).*

In the analysis of detection techniques where growth of microorganisms in a particular medium is required, there are two broad groups of critical control points for microbiological sampling found in the brewing process (*8*). The first group consists of locations where brewing yeast reside and therefore points where selective conditions (i.e. excluding the brewing yeast strain being employed) are required to detect contaminants. Examples of such critical control points include yeast slurry, beer fermentation and aging. The second group consists of locations where brewing yeasts and any other organisms should be either completely absent or present in very low numbers. These critical control points would include raw materials, bright beer, finished product and strategic surfaces of process machinery such as filler heads. In the latter group, relatively non-selective conditions would be desired in order to detect as broad a range of contaminating organisms as possible.

Traditionally, methods for detecting, identifying and characterizing brewery microorganisms, whether at the genus, species or strain level, most often begin by testing for the presence of undesirable microorganisms (i.e. detection) by streaking or inoculating sample to a specific medium (of a selective or non-selective nature), followed by incubation of the medium under defined conditions of temperature and environment (aerobic, anaerobic, or CO_2 enriched) for a lengthy but specified period of time. Detection implies establishing the presence or absence of particular microbes. If further characterization or identification is required, the microbes usually have to be isolated in a pure form, which means restreaking or reinoculating the cells and reincubating them until pure lineages have been isolated, most often as visually discernible colony forming units (CFU). Once purified, specific qualitation or identification of the microorganisms can be conducted through a battery of biochemical (nutrient assimilation), morphological (microscopic) and physiological (e.g. selective staining) tests. While this methodology usually results in accumulation of the required information, it can tend to be too slow and labor-intense to benefit the brewer in time to implement corrective measures. Hence, improvements in microbial detection and identification procedures would be most useful to the brewer. However, an important caveat, described by Barney and Kot (*9*), which must always be taken into consideration when performing 'rapid' analyses of any type which require a growth period for microbes to reach a particular titer, is that microorganisms have their own specific growth rates which may be very slow compared to common clinical species such as *Escherichia coli.* While a rapid detection test for *E. coli,* with a mean generation time of 20-30 minutes, might require only a few hours incubation, the same rapid test for fastidious growers such as *Pediococcus damnosus* or *Lactobacillus lindneri,* with doubling times in excess of five hours, could require several days.

With the relatively recent advances in molecular biology and genetic biotechnology, many techniques developed to characterize organisms at the molecular level and to study their genetic expression under a multitude of conditions have been applied to clinical, food and beverage, and environmental fields in order to achieve one or both of the following goals:
1. Detection of problematic organisms; and/or
2. Identification and/or differentiation of organisms at the genus, species, or subspecies level.

In this chapter, several techniques will be briefly highlighted with suitable references offered for further reading. It is not within the scope of this chapter to discuss all of the different detection and identification methods presently in use, as several lengthy volumes would no doubt be required. The detection methods which will be reviewed in this chapter are: i) impedimetric techniques (conductance, capacitance); and ii) ATP bioluminescence. The identification/differentiation methods which will be reviewed in this chapter are: i) protein fingerprinting by PAGE; ii) immunoanalysis; iii) hybridization using DNA probes; iv) karyotyping (chromosome fingerprinting); and v) polymerase chain reaction (PCR) and RAPD-PCR.

Impedimetric Techniques (Conductance, Capacitance)

The design of impedimetric techniques to detect the presence of microorganisms was based on the observation that growth-related microbial activity could be measured in culture media over time by monitoring changes in the chemical and ionic composition in the medium (*8-20*). When subjecting a substance (such as a growth medium) to an alternating electrical current, the impedance, or resistance to the flow of the current through the medium, is affected by the conductance, or ability of the medium to allow electricity to pass through it, and the capacitance, or the ability to store an electric charge. Using a pair of electrodes submerged in a conducting culture medium, total impedance is the vectorial sum of conductance, which is associated with changes in the bulk ionic medium, and capacitance, which is associated with changes in close proximity to the electrodes (*16*). In a growth medium, large molecules such as carbohydrates are devoid of electrical charge and as a result, increase the impedance of the medium while decreasing both the conductance and capacitance. However, as these uncharged macromolecules are metabolically broken down by microorganisms into smaller subunits such as bicarbonate (which carries a charge), the capacitance and conductance begin to increase while the impedance is diminished. Hence, microbial growth can be monitored in the culture medium by measuring either decreases in total impedance, or increases in either conductance or capacitance.

Two established electrometric instruments that have been on the market for several years are the Malthus 2000 Microbiological Analyzer and the Vitek Bactometer. The Malthus 2000 measures conductance, while the Bactometer measures impedance, or in later models capacitance or conductance. When monitoring a culture using the appropriate medium, measurements of either impedance or conductance, when plotted against time, will produce a curve similar, but not superimposable, to the respective growth curve. Impedimetric measurements require growth of microorganisms to a particular threshold value, which when reached can be accurately detected and measured by computerized accessories. For these instruments, threshold levels, or detection times (i.e. the time required to reach the threshold level in which a change in the electrical property occurs), usually correspond to microbial levels of about 10^5-10^6 CFU/mL. Since detection times are calculated when threshold levels are reached, two factors will significantly affect the detection times of healthy cells inoculated into a suitable medium: i) the quantity of cells initially inoculated; and ii) the natural growth rates or generation times of the respective cultures. In other words, the greater the quantity of viable microorganisms placed initially into the specialized sample cells or the shorter their doubling times, the shorter the detection time. A hypothetical growth and conductance response of two cultures inoculated to the same initial concentration, but possessing different growth rates, is given in Figure 1. The advantages of using impedimetric technology are: i) a time reduction of days in detecting even the most fastidious of microorganisms when compared to standard plating; ii) some selectivity for microbial growth depending on the medium employed; and iii) user-friendly automation for assessing multiple samples for threshold detection. The disadvantages are: i) the systems are expensive to purchase and run; ii) not all types of media are amenable to electrometric analysis due to ionic interferences; and iii) threshold detection levels are typically far in excess of typical brewery plate count standards, which usually fall in the 10^0-10^2 CFU/mL range (depending on the sample location).

ATP Bioluminescence

As with impedimetric techniques, ATP bioluminescence offers the opportunity to significantly reduce, from days to hours, the time required to identify the presence of living organisms. This technology is based on detection of the presence of ATP (adenosine 5' triphosphate) in samples using enzyme-driven light production, or bioluminescence. ATP is a high energy molecule that is found in all living organisms. This molecule can be assayed efficiently by employing the luciferin-luciferase enzyme reaction, which is the basis of bioluminescence in fireflies, by a two-step reaction:

1. Luciferin + Luciferase + ATP + Mg^{2+} $\Rightarrow$
    (Luciferin-Luciferase-AMP) + Pyrophosphate
2. (Luciferin-Luciferase-AMP) + O_2 $\Rightarrow$
    Oxyluciferin + Luciferase + CO_2 + AMP + Light

The amount of light produced correlates with the amount of ATP in the sample. When properly assayed, most (but not necessarily all) ATP is directly indicative of the presence of living organisms in the sample. Commercially available reagent kits are capable of detecting as low as 100 yeast cells per sample without any enrichment, while the respective instruments (i.e. the luminometers) have limits of detection as low as 1-2 yeast cells per sample. Thus, as low as one yeast cell in a given sample with a doubling time of 2-3 hours can be detected after approximately 20 hours incubation. The result is detection of living microbes in considerably less time than required by standard plating. This technique is also comparable in cost to standard plating when factors such as labor and use of disposables such as anaerobic gas paks (when screening for microaerophilic or anaerobic microorganisms) are included.

The use of ATP-driven bioluminescence for the detection of living microbes in non-pasteurized beer at locations along the brew path has become a useful and practical procedure in the brewing industry (*9,11,21-29*). In the last few years, improvements have been made through the development of an efficient reagent for the extraction of ATP from brewery microorganisms (*30,31*) coupled to the development of sensitive luminometers. A recent survey indicated that more than 90 luminometers from more than 60 companies are commercially available (*32*), and some of the more widely used reagents have undergone comparative evaluations (*33*).

ATP bioluminescence, as a quality control device in a brewery, is usually incorporated into hygienic surface swabbing of process machinery, water analysis or beer analysis. As outlined in Figure 2, assays can be designed to detect the presence of total microbial ATP or total non-microbial ATP, with the result being qualitative, but not necessarily quantitative levels of detection compared to numbers of organisms observed by plating analysis, unless pure cultures of organisms are assayed. One reason for this is because different organisms contain vastly different levels of ATP at different phases of growth on a per cell basis. As an example, a typical yeast cell contains approximately 100 times the amount of ATP found in a bacterial cell (*24*). In turn, bacterial cells which vary significantly in size (such as an isolate of *Lactobacillus plantarum* compared to that of *Pediococcus damnosus*) will also vary in relative amounts of ATP per cell. As well, particularly when conducting surface hygiene or water analyses, not all organisms extracted for ATP necessarily grow to colonies when plated on the comparative solid media. Therefore, the employment of bioluminescence for the routine detection of either microbial or non-microbial ATP is useful as a qualitative, or semi-quantitative assay along the brew path.

Protein Fingerprinting by PAGE

Microorganisms generally possess a plethora of two groups of gene products: i) constitutively synthesized structural or regulatory 'housekeeping' gene proteins; and ii) differentially regulated polypeptides that are either induced or repressed as a result of environmental stimuli. While the former category of proteins tend to be found under most conditions conducive to cell viability and growth, the latter category is regulated by conditions that include the medium on which the microorganisms are grown, the incubation temperature, the nature of the gaseous environment, and the growth phase at the time of harvesting. Based on these observations, polyacrylamide gel electrophoresis (PAGE) of cellular proteins has been used in recent years as a means of classifying and identifying yeasts and other microorganisms (*34-41*). In a series of steps, soluble proteins are extracted from pregrown cells (quantities as low as single colony isolates) and subjected to electrophoretic separation. This separation can be made in one dimension based on differences among proteins in size using sodium dodecyl sulphate (SDS PAGE) or native PAGE, or based on differences in ionic charge using iso-electric focusing. Separations can also be performed in two dimensions by combining size and ionic

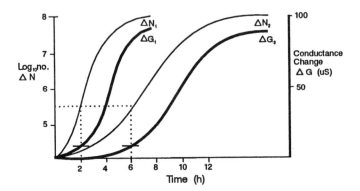

Figure 1. Hypothetical growth and conductance response of two cultures inoculated to the same initial concentration but possessing different growth rates.

Step	A	B	C
1	Microbial & non-microbial cells ⇓	Microbial & non-microbial cells ⇓	Microbial cells
2	Selective release of ATP from non-microbial cells. Microbial cells remain intact	Selective release of ATP from non-microbial cells. Microbial cells remain intact ⇓	
3		Hydrolysis of non-microbial ATP with ATPase. Microbial cells remain intact ⇓	
4		ATPase inactivated. Selective release of microbial ATP ⇓	Selective release of microbial ATP ⇓
5	Addition of luciferin-luciferase	Addition of luciferin-luciferase	Addition of luciferin-luciferase
	⇓	⇓	⇓
6	Luminometer readout (total non-microbial RLU)	Luminometer readout (total microbial RLU)	Luminometer readout (total microbial RLU)

Figure 2. Measurement of ATP-driven bioluminescence. A. Total non-microbial bioluminescence from a mixture of microbial and non-microbial cells. B. Total microbial bioluminescence from a mixture of microbial and non-microbial cells. C. Total bioluminescence from microbial cells only. (RLU = Relative Light Units)

charge separation in one gel. The resulting separation patterns or 'protein fingerprints' are specific to the gene expression of the isolates in question. As such they can be analyzed for relative differences or similarities, and based on this information, the microorganisms can be categorized. The patterns may be compared by visual analysis or densitometer/computer analysis can be used to numerically cluster closely related strains displaying only minor but reproducible differences.

If attempts are made to determine whether or not two or more isolates are essentially the same, it becomes very important to ensure that the respective isolates are treated in the same manner. Since the cellular inventory (and respective quantities) of gene products can be significantly altered by differences in growth conditions, attempts made to compare the protein profiles of one organism to another should include common pregrowth steps immediately prior to harvesting and preparation of cells. Once pregrowth of cells has been completed, extraction, electrophoretic separation and staining of proteins can be accomplished within the span of a few hours.

Immunoanalysis

The use of immunoanalysis has for quite some time been considered as a useful means of identifying contaminating microorganisms along the brew path because of its potential to either detect or identify microbes in either a semi-quantitative or directly quantitative way (*42-46*). Using either polyclonal or monoclonal antibodies, different assays have been designed to differentiate or even speciate microorganisms from each other.

Rapid, semi-quantitative 'sandwich' immunoassays can be used to colorimetrically differentiate or identify relatively abundant ($>10^4$) numbers of intact cells. Such assays employ secondary antibodies to which enzymes have been attached (e.g. alkaline phosphatase, horseradish peroxidase). An example of a sandwich enzyme-linked immunosorbent assay, or ELISA, is given in Figure 3. Typical ELISAs employ a matrix such as a filter membrane, dipstick, test tube or microtitre plate coated with antibody specific to the desired antigen of interest. Sample is placed in contact with the immobilized antibody, allowing binding of antigen to the matrix-bound antibody. The matrix is washed (leaving the bound antigen attached), and a secondary antigen-specific antibody-enzyme conjugate is allowed to bind to matrix-bound antigen. Unbound secondary antibody is thoroughly removed and a substrate is added to the matrix. This substrate is colorimetrically altered by the enzyme attached to the secondary antibody. The result is a semi-quantitative colorimetric change to the matrix if the specific antigen, such as a beer spoilage microorganism, is present. ELISA kits have been designed and are being used extensively as an invaluable tool for environmental testing (for pesticides and other compounds), at-home pregnancy testing, workplace drug screening and even AIDS testing. In the food and beverage industry, problematic pathogens such as *Salmonella* and *Listeria* can be screened with ELISA kits. However, while research is proceeding (*47*), such kits have yet to become commercially available for beer spoilage microorganisms. When available, such kits will be most useful due to their low cost, speed and portability.

Antibodies have been designed and employed to identify brewery microorganisms (48-53). In these cases, single or very low numbers of cells (including microcolonies), have been visualized by immunofluorescent microscopy. Typically, specific antibodies are allowed to bind to contaminant microbes. After removal of unbound specific or primary antibodies, secondary 'indicator' antibodies (such as antibodies specific to the type of immunoglobulin from which the primary antibody is derived) to which fluorochromes (e.g. fluoroscein thiocyanate) have been attached, are added. Unbound secondary antibodies are removed, and microscopic observation of samples under ultraviolet illumination identify the presence of target microorganisms as brightly fluorescing cells.

As with any technique requiring microscopy, eye fatigue can become a problem. As well, it is important to select antibodies that are specific to easily accessible (usually surface) cellular components that tend to be abundantly present regardless of the growth conditions from which the cells were taken, i.e. antibodies raised against cytosolic glucose-repressed or stress-induced gene products would be of limited use. However, as with the use of any specific probe, once potential problems with cross-reactivity of the antibodies to other sources have been resolved, this technique offers fast, qualitative and quantitative analysis of brewery samples without requiring incubation periods for microbial growth to obtain threshold detection values.

Hybridization Using DNA Probes

Deoxyribonucleic acid, or DNA, is the genetic material found in almost all living organisms, except for certain viruses in which the genetic material is ribonucleic acid, or RNA. The biological properties which essentially define all organisms are manifest in sequences of nucleotides, which are made of one of only four bases: adenine (A), cytosine (C), guanine (G), and either thymine (T - found in DNA) or uracil (U - found in RNA). The nucleotide sequences found in different organisms are highly specific and tend to remain for the most part constant from generation to generation. Given the magnitude of roughly millions of bases of DNA found in a typical yeast or bacterium, this detection/identification technique is capable of readily distinguishing specific sequences of anywhere from less than 100 nucleotides to several thousand nucleotides within total genomes.

Hybridization (54) is described as the formation of a double-stranded nucleic acid (either DNA to DNA or DNA to RNA) by base-pairing between single-stranded nucleic acids derived (usually) from different sources. DNA-DNA hybridization and DNA-RNA hybridization are techniques which were developed and utilized primarily by molecular biologists isolating, characterizing and studying the expression of genes (DNA) and gene transcripts (RNA). Of particular interest when performing hybridizations is the ability to control the level of specificity of base-pairing by altering conditions of the hybridization, such as temperature or salt concentrations. When conditions are used to maximize the stringency of the reaction, hybridization becomes a highly selective tool to either differentiate minute differences in DNA sequences that can be associated within and between species of organisms, or to identify target sequences found specifically within the genetic information of certain organisms. The DNA probes employed are sensitive (10^4-10^7 organisms/test sample) and few problems with cross-reactivity are encountered once an identification system has been optimized. With any DNA-DNA hybridization assay, detection of the target organism is not dependent on products of gene expression (e.g. as in PAGE or immunoanalysis) which, as already discussed, can vary with growth conditions, physiological state of the organisms in the test sample, etc.

Hybridization assays can be performed on: i) single colony isolates; ii) cells collected on membrane filters; or iii) purified nucleic acid digested with restriction endonucleases and size-separated by gel electrophoresis (i.e. restriction fragment length polymorphisms or RFLPs). As outlined in Figure 4, diagnostic assays based on DNA hybridization comprise five key components:
1. A means of propagating the organism from the test sample to sufficient titre;
2. A method to release DNA from the target organism;
3. DNA probes specific for the organism of interest;
4. A hybridization format; and
5. A method for labeling of the DNA probes and detecting the resultant hybrids.

A variety of DNA probes have been elucidated and used to identify, differentiate or characterize yeasts and bacteria relevant to the brewing industry. These probes include genes derived from specific microorganisms (e.g. the *HIS4* or *LEU2* genes from *Saccharomyces cerevisiae* or the S-layer protein gene from *Lactobacillus brevis*), endogenous plasmids (such as those found in *Pediococcus damnosus*), transposable or ty elements, arbitrary repeated sequences such as poly GT, or even viral DNA such as the single-stranded phage M13 (*52,55-69*). While this technology has not yet become user-friendly for the brewing industry as a routine quality control test, the potential to produce hybridization kits does now exist. Using hybridization techniques and DNA probes, several commercial diagnostic kits are available which are designed to detect microorganisms (such as *Salmonella* or *Listeria*) with high specificity.

Karyotyping (Chromosome Fingerprinting)

Karyotyping is the determination of chromosomal size and number. Karyotyping has been used for many years to either characterize, differentiate or identify eukaryotic organisms. In higher eukaryotes such as animal and plant species, karyotyping can easily be performed by selectively staining DNA *in situ* (e.g. by use of the Feulgen reaction) and then viewing the clearly discernible chromosomes under a light microscope. The individual chromosomes can then be sorted by micromanipulation and identified. In lower eukaryotes such as yeast, the chromosomes, while still being considered as very large molecules, are comparatively much smaller in size and as such, cannot be readily karyotyped by selective staining and light microscopy. In yeast, karyotyping can be achieved by the electrophoretic separation of whole chromosomes through an agarose gel. Unfortunately, conventional one-dimensional agarose gel electrophoresis, in which DNA fragments approximately 1/100 to 1/1000 the size of a yeast chromosome move through a uniform electric field, cannot resolve large DNA molecules such as

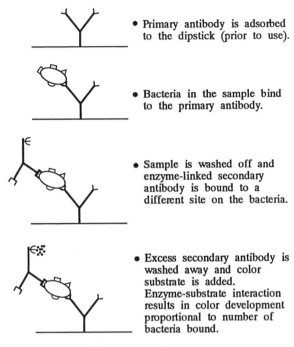

Figure 3. Enzyme linked immunosorbant assay.

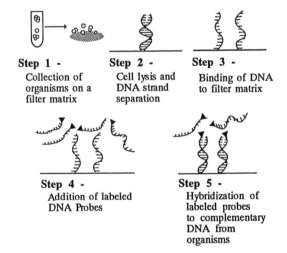

Figure 4. Theoretical DNA-DNA hybridization test for the capture and detection/identification of specific microbes.

yeast chromosomes, since such large molecules tend to migrate independently of their size. However, application of two orientations of electric field has been successfully employed to karyotype yeast chromosomes as a result of the ability of small chromosomes to respond more quickly to changes in the electric field than larger chromosomes. Such changes in electric field to size-separate large DNA molecules are the basis of pulsed-field gel electrophoresis (PFGE).

Chromosomes resolved by PFGE, when stained with ethidium bromide, appear as discrete bands when viewed under ultraviolet light. Various methods for chromosome fingerprinting are available, with the significant difference between them being the orientation of the alternating fields during electrophoresis (Figure 5). As with protein fingerprinting, computer designed cluster analyses can be used to differentiate and/or speciate sample yeast isolates.

Karyotyping of *Saccharomyces* chromosomes has been well documented (*70-76*), and has been utilized in the brewing industry not only as a research and development tool, but from a quality control standpoint as a means of differentiating or fingerprinting ale and lager production strains. This technique has been exceptionally useful in differentiating closely related lager yeast strains through observed chromosomal polymorphisms, which by classical means of characterization (e.g. giant colony morphology, sporulation, growth characteristics, etc.) have often been difficult or even impossible to differentiate (*74,77*). Such chromosomal polymorphisms are the result of insertions, deletions and translocations of DNA fragments large enough (typically 10 kilobase pairs or greater) to be electrophoretically discerned. Pregrown cells can be prepared and electrophoretically separated by PFGE within 48 hours, although more than one run may be required to optimally separate chromosomes of similar size.

Polymerase Chain Reaction (PCR)

One of the newest and potentially most promising techniques that can be applied to rapidly assess the spoilage potential of brewery microbial contaminants is DNA amplification by the Polymerase Chain Reaction or PCR (*78*). PCR is a widely used *in vitro* method for amplifying very small amounts of selected nucleic acids (DNA or RNA) by several orders of magnitude over a short period of time (hours). PCR has revolutionized DNA technology by allowing virtually any nucleic acid sequence to be simply, quickly and inexpensively generated *in vitro* in relatively great abundance and at a discrete length.

At the molecular level, this procedure consists of repetitive cycles of DNA denaturation, primer annealing, and extension by a highly thermostable DNA polymerase. A schematic of this procedure is given in Figure 6. Two short DNA probes (also called primers) of typically 15 to 25 nucleotides flank the DNA segment to be amplified and are repeatedly heat denatured, hybridized to their complementary sequences, and extended with DNA polymerase. The two primers hybridize to opposite strands of the target sequence, such that synthesis proceeds across the region between the primers, replicating that DNA segment. The product of each PCR cycle is complementary to and capable of binding primers, and so the amount of DNA synthesized is doubled in each successive cycle. The original template DNA can be in a pure form or it can be a very small part of a complex

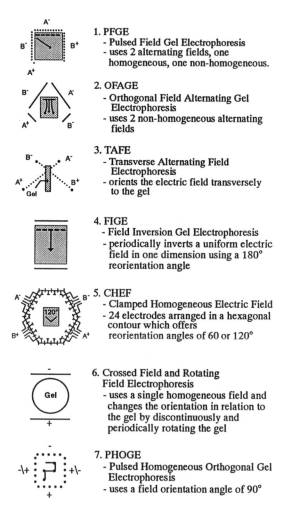

1. PFGE
 - Pulsed Field Gel Electrophoresis
 - uses 2 alternating fields, one
   homogeneous, one non-homogeneous.

2. OFAGE
 - Orthogonal Field Alternating Gel
   Electrophoresis
 - uses 2 non-homogeneous alternating
   fields

3. TAFE
 - Transverse Alternating Field
   Electrophoresis
 - orients the electric field transversely
   to the gel

4. FIGE
 - Field Inversion Gel Electrophoresis
 - periodically inverts a uniform electric
   field in one dimension using a 180°
   reorientation angle

5. CHEF
 - Clamped Homogeneous Electric Field
 - 24 electrodes arranged in a hexagonal
   contour which offers
   reorientation angles of 60 or 120°

6. Crossed Field and Rotating
   Field Electrophoresis
 - uses a single homogeneous field and
   changes the orientation in relation to
   the gel by discontinuously and
   periodically rotating the gel

7. PHOGE
 - Pulsed Homogeneous Orthogonal Gel
   Electrophoresis
 - uses a field orientation angle of 90°

Figure 5. Types of gel electrophoresis systems.

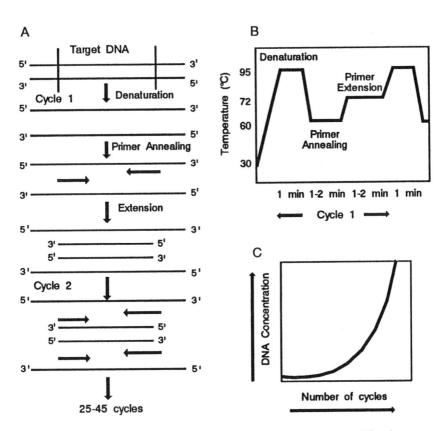

Figure 6. Polymerase chain reaction. A. Schematic of PCR amplification. Each cycle consists of heat denaturation of target DNA, primer annealing, and primer extension with DNA polymerase. B. Time-temperature representation of a typical PCR cycle, which consists of denaturation at 94°C for 1 min, primer annealing at 60°C for 1 to 2 min, and primer extension at 72°C for 1 to 2 min. C. Quantitation of amplified DNA product. Copies of amplified DNA increases exponentially as number of cycles increases.

mixture of biological substances. After 20 replication cycles, the target DNA will (theoretically) have been amplified over a million-fold. Amplified samples can be size-separated by one dimensional agarose or polyacrylamide gel electrophoresis, stained with ethidium bromide and visualized under ultraviolet illuminescence.

To differentiate microorganisms such as yeasts or bacteria at the genus, species, or sub-species level, primers are designed from comparative sequence analysis of variable or highly conserved regions of DNA (rDNA). Different probes can be designed to produce PCR products of various sizes, and the presence or absence of electrophoretically size-separated bands can be indicative of the presence or absence of specific genera or species in the test sample. DNA probes may also be chromogenically 'tagged' to produce PCR products of different colors for differentiation of subsets of microorganisms (79,80). It should be possible to manipulate the specificity of PCR so that, if desired, nucleic acid differentiation could be accomplished for Gram positive or Gram negative bacteria in one scenario, or for *Lactobacillus lindneri* or *Obesumbacterium proteus* in another. Such identification could be elucidated at the single cell level within a yeast slurry, a Helms sample or virtually any sample capable of microbial infestation. To date, although in its relative infancy in the brewing industry, PCR using specific DNA probes has been demonstrated to effectively identify different isolates of brewery microorganisms (81-85).

RAPD - PCR

An alternative to the use of specifically designed probes (i.e. carefully designed to amplify a specific target sequence of 15 to 25 nucleotides) for diagnostic purposes is the use of non-specific or randomly designed probes ranging in size from 9 to 12 nucleotides. The use of these smaller, randomly designed or non-specific probes for PCR diagnostics is called Random Amplified Polymorphic DNA or RAPD. RAPD-PCR will typically produce several PCR products due to binding at many different locations in a genome which can be electrophoretically size-separated to give DNA fingerprints which can be genus or even species-specific (86,87). The RAPD pattern obtained will depend on the sequence of the primers, the reaction conditions for each cycle of amplification and the source of the DNA. This technology has recently been employed to characterize and differentiate a variety of brewery microorganisms including members of *Lactobacillus, Pediococcus, Obesumbacterium* spp. as well as ale and lager yeasts (88,89). In place of a mixture of oligonucleotide probes, RAPD-PCR can be performed using highly repetitive, specific sequences such as poly GT oligonucleotides. Poly GT sequences have been found to occur in the yeast genome at the ends of chromosomes and are believed to be 'hot spots' for genetic recombination (61).

Summary

Improvement and/or development of techniques for detection and identification of microbial contaminants has been significant in the last several years. Elucidation of the presence or absence of microbes can now be performed in only a fraction of the time previously required with classic plating. Rapid verification and identification of the major beer spoilage microorganisms such as *Lactobacillus, Pediococcus, Obesumbacterium,* and other potentially problematic microbes to both the genus and species level constitute major breakthroughs for the brewing industry. Such technology would complement rapid methods presently available which readily detect the presence of living microbes, but lack the ability to satisfactorily differentiate or specify the identity of these microbes, leaving assessment of spoilage potential relatively unknown.

At present, some of the techniques reviewed remain better suited for a Research and Development laboratory rather than a Quality Control laboratory. However, as design of equipment and reagent kits coupled to advancements in methodology catch up to the needs of new customers, these advances in detection and identification will be able to quickly, efficiently and cost-effectively assist the brewer in maintaining beer of the highest quality and consistency through minimization of spoilage brought on by microbial contamination.

Literature Cited

1. Ingledew, W. M. *J. Am. Soc. Brew. Chem.* **1979**, *37*, 145-150.
2. Rainbow, C. In *Brewing Science;* Pollock, J. R. A., Ed.; Academic Press: London, England, 1981, Vol. 2; pp 491-550.
3. Back, W. *Brauwelt Int.* **1987**, *2*, 174-177.
4. Priest, F. G. In *Brewing Microbiology;* Priest, F. G.; and Campbell, I., Eds.; Elsevier Applied Science Publishers Ltd: London, England, 1987; pp 121-154.
5. Van Vuuren, H. J. J. In *Brewing Microbiology;* Priest, F. G.; and Campbell, I., Eds.; Elsevier Applied Science Publishers Ltd: London, England, 1987; pp 155-185.
6. Dowhanick, T. M. *Brew. Dig.* **1990**, *65*, 34.
7. Stenius, V.; Majamaa, E.; Haikara, A.; Henriksson, E.; and Virtanen, H. *Proc. Eur. Brew. Congr.* **1991**, *23*, 529-536.
8. Kyriakides, A. L.; and Thurston, P. A. In *Rapid Microbiological Methods for Foods, Beverages and Pharmaceuticals*; Stannard, C. J.; Petitt, S. B.; and Skinner, F. A., Eds.; Blackwell Scientific Publications: Oxford, England, 1989; pp 101-117.
9. Barney, M.; and Kot, E. *MBAA Tech. Quart.* **1992**, *29*, 91-95.
10. Harrison, J.; and Webb, T. J. B. *J. Inst. Brew.* **1979**, *85*, 231-234.
11. Kilgour, W. J.; and Day, A. *Proc. Eur. Brew. Congr.* **1983**, *19*, 177-184.
12. Firstenberg-Eden, R.; and Eden, G. *Impedance Microbiology*; Research Studies Press Ltd.; John Wiley & Sons Inc.: Toronto, ON, 1984.
13. Evans, H. A. V. *Food Microbiol.* **1985**, *2*,19-22.
14. Adams, M. R. *Anal. Proc.* **1988**, *25*, 325-327.
15. Adams, M. R.; Bryan, J. J.; and Thurston, P. J. *Lett. Appl. Microbiol.* **1989**, *8*, 55-58.
16. Schaertel, B. J.; Tsang, N., and Firstenberg-Eden, R. *Food Microbiol.* **1987**, *4*, 155-163.
17. Unkel, M. *Brauwelt.* **1990**, *4*,100-107.
18. Fleet, G. *Crit. Rev. Biotech.* **1992**, *12*, 1-44.

19. Owens, J. D.; Konirova, L.; and Thomas, D. S. *J. Appl. Bacteriol.* **1992**, *72*, 32-38.
20. Pishawikar, M. S.; Sinhal, R. S.; and Kulkarni, P. R. *Trend. Food Sci. Technol.* **1992**, *3*, 165-169.
21. Avis, J.W.; and Smith, P. In *Rapid Microbiological Methods for Foods, Beverages and Pharmaceuticals*; Stannard, C. J.; Petitt, S. B.; and Skinner, F. A., Eds.; Blackwell Scientific Publications: Oxford, England, 1989; pp 1-11.
22. Dick, E.; Wiedman, R.; Lempart, K.; and Hammes, W. P. *Chem. Mikrobiol. Technol. Lebensm.* **1986**, *10*, 37-41.
23. Harrison, M.; Theaker, P. D.; and Archibald, H. W. *Proc. Congr. Eur. Brew. Conv.* **1987**, *21*, 168-173.
24. Hysert, D. W.; Kovecses, F.; and Morrison, N. M. *J. Am. Soc. Brew. Chem.* **1976**, *34*, 145-150.
25. Krause, J. A.; and Barney, M. C. The use of bioluminescence for the rapid detection of beer spoilage microorganisms; Presented at *The 53rd Annual Meeting of the American Society of Brewing Chemists*: Cincinnati, Ohio, 1987.
26. Miller, R.; and Galston, G. *J. Inst. Brew.* **1989**, *95*, 317-319.
27. Simpson, W. J. *Brew. Guardian.* **1989**, *118*, 20-22.
28. Simpson, W. J.; Hammond, J. R. M.; Thurston, P. A.; and Kyriakides, A. L. *Proc. Congr. Eur. Brew. Conv.* **1989**, *22*, 663-674.
29. Takamura, O.; Nakatani, K.; Taniguchu, T.; and Murakami M. *Proc. Congr. Eur. Brew. Conv.* **1989**, *22*, 148-153.
30. Jago, P. H.; Stanfield, G.; Simpson, W. J.; and Hammond, J. R. M. In: *ATP Luminescence - Rapid Methods in Microbiology.* Stanley, P. E.; McCarthy, B. J.; and Smither, R., Eds.; Blackwell Scientific Publications: Oxford, England, 1989; pp 53-61.
31. Simpson, W. J.; and Hammond, J. R. M. In *ATP Luminescence - Rapid Methods in Microbiology*, Stanley, P. E.; McCarthy, B. J.; and Smither, R., Eds.; Blackwell Scientific Publications: Oxford, England, 1989; pp 45-52.
32. Stanley, P. E. *J. Biolumin. Chemilumin.* **1992**, *7*, 77-108.
33. Griffiths, M. W.; and Phillips, J. D. In *Rapid Microbiological Methods for Foods, Beverages and Pharmaceuticals*; Stannard, C. J.; Petitt, S. B.; and Skinner, F. A., Eds.; Blackwell Scientific Publications: Oxford, England, 1989; pp 13-32.
34. Kersters, K.; and De Ley, J. *J. Gen. Microbiol.* **1975**, *87*, 333-342.
35. Van Vuuren, H. J. J.; Kersters, K.; De Ley, J.; and Toerien, D. F. *J. Appl. Bacteriol.* **1981**, *51*, 51-65.
36. Drawert, F.; and Bednar, J. *J. Agric. Food Chem.* **1983**, *31*, 848-851.
37. Van Vuuren, H. J. J.; and Van Der Meer, L. J. *Am. J. Vitic.* **1987**, *38*, 49-53.
38. Van Vuuren, H. J. J.; and Van Der Meer, L. J. *J. Inst. Brew.* **1988**, *94*, 245-248.
39. Dowhanick, T.; Sobczak, J.; Russell, I.; and Stewart, G. *J. Am. Soc. Brew. Chem.* **1990**, *48*, 75-79.
40. Vancanneyt, M.; Pot, B.; Hennebert, G.; and Kersters, K. *System. Appl. Microbiol.* **1991**, *14*, 23-32.

41. Newsom, I. A.; O'Donnell, D. C.; and McIntosh, J. *Inst. Brew. Proc. Australia, New Zealand Sect. Conv.* **1992**, *22*, 189.
42. Campbell, I.; and Allan, A. M. *J. Inst. Brew.* **1964**, *70*, 316-320.
43. Tsuchiya, T., Fukazawa, Y., and Kawakita, S., *Mycopathol. Mycol. Appl.* **1965**, *26*, 1-15.
44. The Institute of Brewing Analysis Committee. *J.Inst. Brew.* **1973**, *79*, 134-136.
45. Dolezil, L.; and Kirsop, B. H. *J. Inst. Brew.* **1975**, *81*, 281-286.
46. Hutter, K.-J. *Brauwiss.* **1978**, *31*, 278-292.
47. Umesh-Kumar, S.; and Nagarajan, L. *Folia Microbiol.* **1991**, *36*, 305-310.
48. Legrand, M.; and Ramette, P. *Cerevisiae.* **1986**, *4*, 173-182.
49. Phillips, A. P.; and Martin, K. L. *J. Appl. Bacteriol.* **1988**, *64*, 47-55.
50. Yasui, T.; Taguchi, H.; and Kamiya, T. *Proc. Congr. Eur. Brew. Conv.* **1989**, *22*, 190-192.
51. Hutter, H.-J. *Brauwelt Int.* **1992**, *2*, 183-186.
52. Vidgren G.; Palva, I.; Pakkanen, R.; Lounatmaa, K.; and Palva, A. *J. Bacteriol.* **1992**, *174*, 7419-7427.
53. Whiting, M.; Crichlow, M.; Ingledew, W. M.; and Ziola, B. *Appl. Environ. Microbiol.* **1992**, *58*, 713-716.
54. Wetmur, J. G. *Crit. Rev. Biochem. Mol. Biol.* **1991**, *26*, 227-259.
55. Decock, J. P.; and Iserentant, D. *Proc. Congr. Eur. Brew. Conv.* **1985**, *20*, 195-202.
56. Martens, F. B.; Van Den Berg, R.; and Harteveld, P. A. *Proc. Congr. Eur. Brew. Conv.* **1985**, *20*, 211-217.
57. Pedersen, M. B. *Carls. Res. Comm.* **1983**, *48*, 485-503.
58. Pedersen, M. B. *Carls. Res. Comm.* **1985**, *50*, 263-272.
59. Pedersen, M. B. *Carls. Res. Comm.* **1986**, *51*, 163-183.
60. Steffan, R. J.; Breen, A.; Atlas, R. M.; and Sayler, G. S. *Can. J. Microbiol.* **1989**, *35*, 681-685.
61. Walmsley, R. M.; Wikinson, B. M.; and Kong, T. H. *Bio/Technol.* **1989**, *7*, 1168-1170.
62. Delley, M.; Mollet, B.; and Hottinger, H. *Appl. Environ. Microbiol.* **1990**, *56*, 1967-1970.
63. Meaden, P., *J. Inst. Brew.* **1990**, *96*, 195-200.
64. Colmin, C.; Pebay, M.; Simonet, J. M.; and Decaris, B. *FEMS Microbiol. Lett.* **1991**, *81*, 123-128.
65. Lonvaud-Funel, A.; Fremaux, C.; Biteau, N.; and Joyeux, A. *Food Microbiol.* **1991**, *8*, 215-222.
66. Lonvaud-Funel, A.; Joyeux, A.; and Ledoux, O. *J. Appl. Bacteriol.* **1991**, *71*, 501-508.
67. Miteva, V. I.; Abadjieva, A. N.; and Stefanova, Tz.T. *J. Appl. Bacteriol.* **1992**, *73*, 349-354.
68. Reyes-Gavilan, C. G. de los; Limsowtin, K. Y.; Tailliez, P.; Sechaud, L.; and Accolas, J.-P. *Appl. Environ. Microbiol.* **1992**, *58*, 3429-3432.
69. Lonvaud-Funel, A.; Guilloux, Y.; and Joyeux, A. *J. Appl. Bacteriol.* **1993**, *74*, 41-47.

70. De Jong, P.; De Jong, C. M.; Meijers, R.; Steemsa, R.; and Scheffers, W. A. *Yeast.* **1986,** *2,* 193-204.

71. Johnston, J. R.; and Mortimer, R. K. *Int. J. System. Bacteriol.* **1986,** *36,* 569-572.

72. Casey, G. P.; Xiao, W.; and Rank, G. H. *J. Inst. Brew.* **1988,** *94,* 239-243.

73. Degre, R.; Thomas, D. Y.; Ash, J.; Mailhiot, K.; Morin, K.; and Dubord, C. *Am. J. Enol. Vitic.* **1989,** *40,* 309-315.

74. Casey, G. P.; Pringle, A. T.; and Erdmann, P. A. *J. Am. Soc. Brew. Chem.* **1990,** *48,* 100-106.

75. Vezinhet, F.; Blondin, B.; and Hallet, J. N. *Appl. Microbiol. Biotechnol.* **1990,** *32,* 568-571.

76. Querol, A.; Barrio, E.; and Ramon, D. *System. Appl. Microbiol.* **1992,** 15, 439-446.

77. Good, L.; Dowhanick, T. M.; Ernandes, J. E.; Russell, I.; and Stewart, G. G. *J. Am. Soc. Brew. Chem.* **1993,** *51,* 35-39.

78. Mullis, K. B.; Faloona, S. J.; Scharf, S. J.; Saiki, R. K.; Horn, G. T.; and Erlich, H. A. *Cold Spring Harbor Symp. Quant. Biol.* **1986,** *51,* 263-273.

79. Embury, S. H.; Kropp, G. L.; Stanton, T. C.; Warren, T. C.; Cornett, P. A.; and Chehab, F. F. *Blood.* **1990,** *76,* 619-623.

80. Kropp, G. L.; Fucharoen, S.; and Embury, S. H. *Blood.* **1991,** *78,* 26-29.

81. Klijn, N.; Weerkamp, A. H.; and de Vos, W. M. *Appl. Environ. Microbiol.* **1991,** *57,* 3390-3393.

82. Tsuchiya, Y.; Kaneda, H.; Kano, Y.; and Koshino, S. *J. Am. Soc. Brew. Chem.* **1992,** *50,* 64-67.

83. Tsuchiya, Y.; Kano, Y.; and Koshino, S. *J. Am. Soc. Brew. Chem.* **1993a,** *51,* 40-41.

84. Tsuchiya, Y.; Kano, Y.; and Koshino, S. Identification of lactic acid bacteria using temperature gradient gel electrophoresis after rapid detection by polymerase chain reaction; Presented at *The 59th Annual Meeting of the American Society of Brewing Chemists*: Tucson, Arizona, 1993b.

85. DiMichele, L. J.; and Lewis, M. J. Rapid, species-specific detection of lactic acid bacteria from beer, using the polymerase chain reaction; Presented at *The Second Brewing Congress of the Americas*: St. Louis, Missouri, 1992.

86. Williams, J. G. K.; Kubelik, A. R.; Livak, K. J.; Rafolski, J. A.; and Tingey, S. V. *Nucl. Acid. Res.* **1991,** *18,* 6531-6535.

87. Waugh, R.; and Powell, W. *Trend. Biotech.* **1992,** *10,* 186-191.

88. Savard, L.; and Dowhanick, T. M. Polymerase chain reaction using arbitrary primers for rapid characterization of brewery microorganisms; Presented at *The 59th Annual Meeting of the American Society of Brewing Chemists*: Tucson, Arizona, 1993.

89. Savard, L.; Hutchinson, J.; and Dowhanick, T. M. Characterization of different isolates of *Obesumbacterium proteus* using random amplified polymorphic DNA; Presented at *The 59th Annual Meeting of the American Society of Brewing Chemists*: Tucson, Arizona, 1993.

RECEIVED March 30, 1993

Sensory Characterization and Control

Chapter 3

Sensory Science

A Brief Review of Principles

Christina W. Nasrawi

Monsanto Agricultural Company, 700 Chesterfield Parkway North, St. Louis, MO 63198

A brief review of principles, applications, and guideline for conducting research in sensory science and using sensory methods to achieve test objectives and research goal.

Input from sensory receptors such as visual, auditory, olfactory, gustatory and tactile origins allows perception or detection of color, dimension, sounds, odor, taste, texture and other tactile stimuli that cause irritation, pain, pressure, temperature etc. (*1*). Most of these inputs are taken for granted until such time of a loss. Marketing specialists understand the importance of human perception. To promote sales at the market place, products are often displayed in manners that optimize products' sensory appeal ultimately to promote interests, sales and utilization (*2-4*). Studies have demonstrated the viable relationships between sensory input and behavioral pattern of food selection, acceptance, and preference (*5,6*); and intake (*7-9*). It can be postulated that the mechanism underlying the process of selection and consumption of food and non-food products, from soda through clothing to lawn-mowers, is of sensory origin.

Sensory science draws information from a composite of scientific areas that ranges from anatomy and physiology, through foods and nutrition, to psychophysics and psychology. Therefore, it is impossible to cover the entire topic of sensory science in great detail in a chapter. This chapter is written in an attempt to provide the lay and scientific communities with a general overview on sensory science and its methods. The very basic of principles and guidelines on the techniques of sensory evaluation will be discussed, to give the readers a sense of what, when, where and how to use (or not to use) the various sensory evaluation methods. Readers who require more detail within a specific area of sensory science, should consult available publications, journals, and technical manuals in the specific area of interest.

"A Scientific Discipline"

It can be said that sensory science is "as old as Noah". Utilization and trading of spices and dyes are well documented as part of the development of human civilization. Spices were valuable and widely used as flavor enhancing ingredients and for food preservation. Several New Testament books in the Bible tells of gifts (frankincense

0097–6156/93/0536–0032$06.00/0

and myrrh) brought by the wise men to the infant Christ. Leaping forward in time, peer recognition of sensory science as a viable scientific field, did not take place until the World War II era. Then, it was the food, beverages and personal care products industries that generated most of the interests in sensory science for one obvious reason - to promote sales and consumption. Reports of sensory evaluation of perfume, spices, coffee, tea, spirits, and wine are well documented (*10-18*).

The term "organoleptic" test, roughly translates to mean evaluation with "organs", was widely used in most of the earlier sensory works. Most of these works were found to lack a define set of principles, standardization of nomenclature, techniques, and direction (*19*). The progression from organoleptic tests to sensory science was led by a list of scientists inclusive of Little, Ellis, Foster, Kramer, Dawson, Amerine, Krum, Mitchell, Schwartz, Peryam, Pangborn, and a team of statisticians. Resulting from their combined efforts, in 1957, the first symposia on sensory evaluation methods was held by the Institute of Food Technologists (IFT), and in 1961, a technical committee, E-18, on sensory evaluation of materials and products was formed in American Society for Testing and Materials (ASTM) (*19*).

Presently, due to world wide expansion of interests, research contributions from scientists of diverse fields, and inclusion of sensory programs into Food Science curricula at both undergraduate and graduate levels, sensory science has come of age. The future lies with continuing interests and recognition, at the academic, research, and industry levels, for its practical aspect in elucidating the relationship between perception and human response.

"Sensory or Analytical Instruments"

Historically, invention of machines and instruments have served in improving the quality of life. Mechanical devices were most effective in saving time, energy, resources, and quite often, human lives. With the explosion of technology, and accelerated pace of information exchange in both the natural and applied sciences, the functional capabilities of machines are being upgraded to levels of sophistication that were not in existence only a few decades ago. In the laboratory setting, instruments allow researcher to meet set objectives and goals without spending valuable resources on tasks which are mundane and repetitive in nature and perhaps most important of all, allow research scientists to go home at night.

However, the cost of such instrument are often exorbitant, thus savings from the financial standpoint can be questionable. Furthermore, the cost factor often poses a limitation on the availability of such instrumentation in institutions with lesser monetary resources. Nevertheless, machines do not tire, nor acquire motivational problems as humans do, and provide tangible proof of work. A quick glance through the various equipment catalogs would reveal the primary function of most instruments. They are designed for a specific task, such as to bend, separate, heat, condense, reduce, pressurize, clean, digest, and protect. It is true, without these instruments, research projects would grind down to a slow crawl, with little progress in sight.

It is apparent that sensory methods would not provide information at the molecular, atomic and sub-atomic levels; exact concentration of solutes, chemical reactions, cellular kinetics and events. But it will provide information on whether events at cellular, sub-cellular, molecular, atomic, and sub-atomic levels are perceivable by human senses. It takes human senses to provide information on human perception. Instruments can mimic but will never provide true human response.

"Subjective or Objective"

Argument for subjectivity of human response stems from difference between individual. It is true, no one person is identical to the next, not even identical twins. Based on this viewpoint, human responses would be subjective, in comparison to instruments that generate objective measurements (1). When it comes to sensory information, it is ironic that many scientist and lay people tend to accept numbers and figures that are generated by instruments faster than what they perceive through their senses. It is important to note that instruments are operated by human and/or computers, which are made, programmed, and calibrated by humans. Data acquired are grossly dependent upon the operator, and is further complicated by the many variation of methods and techniques.

Nevertheless, the objectivity of sensory methods; positive correlation between instrumental and sensory data; and the ability of human olfactory receptor to perceive odorous compounds at lower concentration levels than the most sensitive chromatographic detector, are well documented (20-27). Correlations between sensory and instrumental data were established to better the use of analytical instruments in predictions of perception.

The objectivity of certain sensory methods was best demonstrated by Stevens et al (28) where the authors established positive correlations between sensory and instrumental data. Following the principle of chromatography, GC separates compounds, qualitatively and quantitatively, and displays compounds as peaks on chromatogram. However without known standard compounds, information on the chromatogram is quite meaningless. Even with known standards, chromatogram does not provide information on the aroma characteristic of each peak (29).

Stevens et al., (30) split the GC effluent port, thus allowed simultaneous sniffing and detection of the separated compounds by human subjects and detector respectively. Aroma descriptors generated by test subjects resulted in the correlation of the vine-like or green aromas with hexanal and hex-trans-2-enal, and the fruity or floral aromas with 6-methyl-5-hepten-2-one and -ionone.

Wood (31) optimized tomatoes flavor with the GC-sensory characterization technique. In addition, the GC-sensory characterization technique has been used in conjunction with time-intensity module to characterize the temporal component of GC effluent compounds (32). Far from being subjective, sensory evaluation methods are providing objective data that instruments can not provide at any cost.

"Sensory in the Big Picture"

Movement towards a market economy on a global basis, is truly evident. To reap financial rewards, industries must supply consumers with products and services, that satisfy consumers' expectation. Thus the primary goal of industries is to invent, make, package, and sell products that would meet the needs and/or demand of consumers. To supply a qualified work force, academia must educate and train students, to meet industries' job-demand. At the cutting edge, research institution, academically linked or not, must gather and provide information.

As part of this global machinery, sensory science incorporate information on human perception and response. Within a product industry, sensory information is useful in: quality standards (USDA-US Standards); shelf-life and changes during transportation and storage (33,34); quality control, improvement, and assurance (4,35-38); product research and development (39-44); and management (45).

Gatchalian et al., (46) modified the quality control circle of Kramer et al., (47,48) illustrating the role of sensory evaluation in relation to activities within a product industry. It is necessary to add academic and research institutions into the scheme of information transfer. Sensory evaluation is strategically placed in the center of all activities, where it exchanges information with the next concentric ring (profiling of products, development of test procedures, conduct tests, reporting of results, trouble shooting, and determination of consumer trends), which in turn exchanges with the outer circle (sales, research and development, quality control, and production), and so on. It is simplistic in description, but adequately describe the role of sensory evaluation in product industries.

Zelek (*49*) eloquently summarized the importance of conducting sensory work to substantiate claims in advertisements. The author cited numerous examples of legal ramification resulting from failure in substantiation of product claims. The author further suggested that sensory evaluation of products be carried out with the specific intent of substantiating manufacturer's claim. The domain of sensory evaluation appears to be expanding as its practical aspects are being realized.

"Some Basic Terms"

Sensory science is defined as the scientific discipline used to evoke, measure, analyze and interpret sensation as they are perceived by the senses of sight, smell, taste, touch, and hearing, by the IFT (*50*). In simple term, it allows research scientists to systematically qualify and quantify human perceptions. Of course, a set of definitions, terminology and nomenclature comes with sensory science. With continued research and development, modification of existing sensory language is a certainty.

The term "tactile" is used to describe textural properties of products, but has also been known as kinesthetic (*51*), and haptaesthesis (*52*). Because of their depictions of muscular responses, the later terms are not frequently found in sensory reports. Confusion between "taste", "aroma", and "flavor" attributes is best cleared by defining flavor. Flavor is defined as, a mingled but unitary experience which includes sensations of taste, smell, pressure, and other cutaneous sensations such as warmth, cold, mild pain (*53*).

Tastes are gustatory origin, and can basically translate to perception of the classical, sweet, sour, salt and bitter tastes. Recent work on flavor enhancers such as monosodium glutamate (MSG) and 5'-nucleotides (IMP, GMP), have led to characterization of taste quality related to these compounds. In addition to the classic four, the term umami is used to describe taste quality associated with flavor enhancers (*54*). Completely independent from tastes, aroma and odor are terms frequently used to describe olfactory sensations or smell. It is impossible to define appearance without such terms as color, lightness, saturation, hue, absorption, reflectance, transparency, translucency, and opacity. Glossary of sensory terms can be found in Amerine et al., (*53*), and other recent publications (*55,56*).

"Research Principles"

Successful completion of an experimental project is measured by its completion of defined objectives. The ultimate goal is to test the hypothesis - the question. In a similar manner, sensory research does not deviate from this protocol. Pangborn (*19*) reviewed the historical development of sensory science, and emphasized the importance of setting clearly defined objectives.

It is unfortunate, that the science (or art) of setting goals, and defining objectives is not easily grasped, and is a difficult point to teach. Quite often, it is left at an intuitive level for researchers (or graduate student) to stumble through the research jargons such as goals, objectives, experimental design, methods, data collecting procedures, data evaluation, interpretation etc. Most sensory texts are guilty in the same sense, where readers are bombarded with descriptions of sensory evaluation methods, along with extensive statistical procedures, with little guidance on what, when and where to use them. Lacking clear guidance, many scientists are "stuck on" difference tests, such as paired comparison, triangle, and duo-trio tests, and hedonic scales, because they are "easiest" to use.

When goals and objectives are clearly set, in many cases, they would prescribe the sensory method. However, there are important criteria that must be established, addressed, and clarified, prior to choosing method for sensory evaluation. Extensive investigation on subject matter should be conducted that include, review of reported literature, communication with product specialists, or researchers experienced within the area of interest. For example, much of the anatomical and physiological information on color (58), taste (59), odor (60), flavor (61) and tactile (1) perceptions, is essential for adequate understanding, interpretation, and discussion of sensory data. As stated earlier sensory measurements integrate information from many different fields, ranging from molecular biology, through psychology, to computer statistics.

In addition to comprehensive literature search, assessments on the availability of test site, test subjects, statistical and computer (soft and hardware) supports must be carried out. Lacking in any one of these components may limit the scope of the research project, affect the quality of data collected., and in worst cases, fail to meet set objectives.

There are numerous publications on the requirement of the physical lay out of sensory facilities (53,61-63), recruiting and training of test subjects (22,23, 64-67), for readers to consult. Test facilities should be isolated from extraneous odor, color, light, noise; have booths for effective isolation of test subjects; and have adequate ventilation, temperature and humidity controls. Additional amenities such as drinking fountains in test booths, fixed drains for expectoration, signal light switches, and computerized data collection systems, could be implemented based on need and availability of funds.

The importance of product knowledge should not be over emphasized. The type of product would dictate sample size; number of sample per test session; if temporal or time related component should be addressed; the test protocol; and the availability, and motivation level of test subjects. Due to the complex compositional nature of food and beverage systems, sensory evaluation of these products underscores the researcher's ability to control variability within experimental parameters.

Information on product history should accompany products brought to sensory laboratory, such as, sampling procedure, point of origin, storage, to ensure that sample and sample size, represents the product. For example, in continuous aseptic processing system, instead of cleaning the line after each product (costly in time and labor), quite often products are processed one right after another, with a small overlapping of products mixture between products (68). Therefore, samples collected that are too close to the previous product may not be considered as representative.

In the laboratory, the first order of event is to run a "bench-top" session with the product. Bench-topping is a common terminology used to describe the preliminary investigation of product. It can be formal or informal, with the end result providing valuable information on the basic characteristic of the product. Information ranges from its physical nature, through differential flavor components, to stability of the product. In most instances, bench top sessions provide researchers with a rough direction on sample handling, sampling, preparation, sizing, portion, receptacle choice, number of samples, light source, sensory method, and testing procedure.

For example, in testing of chili pepper, or the principle component of chili pepper, capsaicin, for mouth burn intensity, the irritative nature of the sample posed restrictions on many of the above stated criteria (69-73). Due to the temporal component of capsaicin's irritation, the time-intensity method provided data showing the onset and decay of mouth burn. The irritative nature placed restriction on number, size, concentration of samples per test session. Test procedure was modified to allow ample amount of time between samples for mouth burn to subside (74).

Akin to other areas of scientific research, it is prudent to consult statistician, for experimental design, subjects requirement, statistical method for data analysis, and availability of computing devises for statistical computations. Furthermore, it would be devastating to discover after the completion of research project, that result lacks statistical bearing, and therefore lacks validity. For conservation of time, energy and resource, it is best to predetermine the availability of computer support, in terms of program, computing devise, and program consultant. The logistic of handling large sets of numbers without the aid of computing device would definitely result in computing error. For example, it is extremely difficult to carry out principle component analysis (PCA) without computing devise, programs, and consultant. Even with the availability of computing devise, care should be taken in entering data, as it is the primary source of error caused by inaccurate data entry.

The number of test subjects required for a given study or research project is best estimated statistically based on the number of variables within a set confidence interval of mean to ensure conservative testing of the null hypothesis for statistical validity of results (75-77). A quick and dirty alternative in determining a statistically wholesome number of subjects would be to use information available in published literatures. It is recommended that a larger than necessary number of subjects be recruited, because for various reasons, some will drop out. In addition, in affective or consumer acceptance studies, it is essential for recruits to represent a certain target population, or segment of population. It is futile to use dogs in testing of rat chow.

"Sensory Methods"

Different methods available are designed for distinctly different types of sensory evaluation, akin to the many types of chemical assays available within the realm of analytical chemistry. Used appropriately, no one method is better or worst than another. Again, much like chemical assays, each method comes with a set of limitation and benefit (53,78,79).

The task at hand is to choose appropriate method(s), that would allow gathering of sufficient quantity of valid data to meet objectives. Points of benefits (advantages) vary from; simple and quick, to reliability and repeatability. Most limitations (disadvantages) originate from the mechanics of the methods, such as length of test session, complicated procedure, and the dept of subject training. Quite often, a quick and simple discrimination method, such as paired comparison, would provide quick answers (sample A is different from sample B), but provide little data on where the differences lie. On the other hand, quantitative method such as the time-intensity method is lengthy and often requires a fair amount of subject training, a hindrance, but would quantitatively explain the difference between samples on a temporal basis, which represents an advantage.

There are many workshops and short-courses organized by consultant firms, and various educational institutions; and symposiums (ASTM, IFT, AChemS etc., and other foods and nutrition professional associations of United States and European origin), that emphasize specific areas of sensory evaluation, such as consumer tests, qualitative and quantitative tests, and sensitivity tests, as well as statistical evaluation. Again, readers who are interested in a specific area of sensory evaluation, and for various reasons could not enroll in a sensory evaluation class, are urged to utilize these "hands-on" courses. Furthermore, these short-courses and seminars, often provide valuable information from a technical stand-point, thus, are excellent for application of learned theory.

In a nutshell, Pangborn (80), illustrated the classification of various methods available for sensory evaluation (Figure 1). A test solution, sample or product held at a concentration capable of eliciting a sensory response could be subjected to two classes of tests: (1) Analytical-Laboratory tests; and (2) Consumer tests. At the analytical level, the above mention test solution could be subjected to: (1) Sensitivity; (2) Quantitative; and (3) Qualitative tests. Depending on the type of sample, set experimental objectives, and the scope of a given project, it may become essential to utilize a wide range of test methods.

A quick example of sensory evaluations at the analytical level would be, upon the discovery of an odorant compound, it may be essential to determine its detectable threshold concentration; absolute threshold concentration; increasing or decreasing concentration steps that would be noticeable different; determine differential detection concentration between sex, age groups etc.; quantitative scaling to correlate between molar concentrations and perceptual increase in intensity; determine the temporal profile of the odorant at varying dosage from onset to decay; qualitative studies to better

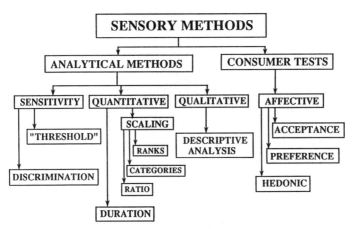

Figure 1. Classification of Sensory Methods (reproduced with permission from Cereal Foods World, vol. 25: 10. Copyright 1980 American Assoc. Cereal Chem.).

describe the aromatic qualities at different concentrations, in different media, and mode of delivery or application; identify differential qualities between isomeric compounds; and compare qualitative profile of the odorant to existing products in the market.

"Functions of Methods"

Consumer Tests. Puzzled on which of the methods would effectively accomplishes the task in question? The first round of elimination would be whether consumer oriented test and survey are of interest. Consumer tests provide product industries with essential information on performance potential of the product in the market (*4,81*). Consumer tests are designed to measure consumer response on a target product, to test consumer acceptance of a product. Therefore to determine whether consumers like-dislike, accept-reject, and preference of the product in question (*53,82*). It is important to take into account factors of non-sensory origin, such as nutrition; mode of utilization; behavioral and context effect. For example, within a reasonable price range, and given a choice between product A in glass containers versus product B in tin cans, consumers may choose product A over B because glass is being collected and recycled. He may not buy the product again if its sensory properties fall below his acceptance standards, but his initial purchase decision has stemmed from concerns over the environment and the availability of neighborhood recycling program.

Consumer tests can be conducted "in-house" and "on locale". Taking the test to a central location, requires a large pool of test subjects to provide proper representation of the target population (sex, age group, socio-economic status, ethnic background, etc.), thus the logistics of distributing samples can be laborious and costly. "In-house" laboratory consumer test utilizing company employees for test subjects, can be used as an effective indicator of "the real market" and should be carried out prior to large scale testing at the market place. The advantages of conducting in-house testing include the availability of controlled conditions, rapid data feed-back, abundant supply of subjects acquainted with test procedure, and economy. Disadvantages include, subjects' familiarity with the products, and inappropriate representation of target population sector.

When large scale consumer testing is eminent, selection of test sites that match the demograph of target population follows. As stated earlier, screening of large number of subjects that match the demographic profile is essential for proper representation. The extent of geographic coverage, and mode of test products distribution would depend on the availability of technical support, facilities, and funds.

Tests may be conducted at a fixed central location, a mobile laboratory, or "in-home". Moving the test to central location(s) makes available a large number of test subjects, but eradicates the control test environment that in-house laboratory provides. The mobile laboratory set up, to a certain degree, is akin to taking the laboratory to the consumers. It provides a better control of test environment than the central locale set up. But context effect should be accounted for, as subjects are forced to evaluate samples out of context of normal home use.

The third form of testing involve home placement of test products. Home placement allows subjects to utilize or test products under actual condition of use. Through survey forms, tests can be expended to include input from members of subject's family, and marketing information. There is no time restraint set on the subjects as in central locale, and mobile laboratory set ups. However, with home placement or "in-home" consumer tests, there is a complete loss of control, with no guarantee of subject responding or providing the data sought, within set time limit, or ever. Also, for the amount of data generated, the logistics of products distribution can be very expensive.

In addition to determining effective test locales, the rest of the test parameters must be defined. Experimental design would be dependent upon product characteristics. The sensory characteristics of the product would dictate the number of samples, and test procedure. For example, in testing of chili sauces, the futility in testing of multiple samples is obvious. Therefore, the mode of sample presentation along with test procedure would vary. Samples can be presented to test subjects one at a time or all at once. It is important to understand that test subjects are not "test-wise", thus procedure should be kept simple without tasks of bothersome nature, utilizing the simplest of instruction. It is important to note that the choice in test methods, must be coordinated with the target population (age, behavior, cognitive level etc.), facility available at the test locale (space, equipment, flow of traffic etc.), sensory characteristics of the products (aftertaste, interaction between products, etc.), and test objective or the type of affective information.

Regardless of method, test sites, design, etc., the test procedure should be in the easiest and simplest of form, keeping in mind that members of consumer panels are not acquainted with sensory protocol. Complicated procedures would only cause confusion, and carry over into their response, leading to disaster. Generally, products are assessed by directing test subjects to answer question related to products, choose a product, rank, or rate products in accordance to preference. Lastly, statistical manipulation of consumer data should reflect the study design, and thus appropriately, the type of analysis may range from Chi-square distribution, through analysis of variance with mean comparison, to multivariate type analyses.

Ideally, consumer test results should serve as a good indicator for performance of products in market place, by gathering data on relationship between product characteristic and consumers' expectation. With all of its advantages and limitations, consumer tests are the only effective mode of measuring consumer response to products. Even after conducting all the intricate chemical/instrumental proximate analyses and objective sensory tests, without extensive consumer tests, a theoretical "winner" may turn out to be a "dud", because the product does not meet the needs or standards of consumer's expectation.

Sensitivity Tests.   Distinctively different from consumer tests, analytical/laboratory tests are often recognized as in-house objective tests within the product industries, but are the major tools in conducting basic sensory research in educational and research institutions. These methods are designed for specific function, to meet specific test objectives. In general, sensitivity tests are used in relative measurement of strength of compounds, sensitivity of test subjects, and methods.   They are also useful for measuring the relationship between chemical constants and human perception (*83*). It is important to note that sensitivity tests only measure the relative intensity of substrates. However, sensitivity tests are an integral part of qualitative studies such as in descriptive sensory analysis of products (see Qualitative Analysis).

Of the sensitivity tests, threshold evaluation allow measurement of perception at detectable (or absolute), discriminable (or difference), and recognizable (or identify) levels of specific substrates concentration. Simplified, threshold is a statistically-determined endpoint (95% confidence limits) that is part of a dynamic sensory continuum (*83*). A continuum of sensory spectrum exists in any given stimulus and it ranges from the non-perceptible through recognition to the terminal threshold where further increase in concentration are no longer discernable, and instead pain is often invoked. For example, the stimulus concentration at which a transition occurs from no sensation to sensation is designated at absolute or detection threshold. Further up the continuum, just noticeable difference (JND) is the smallest change in the physical intensity of a stimulus which is perceptible. JND test is a good tool to measure the ability of test subjects to differentiate concentration of stimulus. A practical application of JND would be in ingredient (re)formulation, it is important to determine the concentration at which a perceptible difference is noticed as an ingredient concentration is being changed.

It is apparent that sensitivity of subjects would depend a great deal on the type of stimulus, the decision making process, the type of method used in threshold measurements, and the intensity of stimulus (*84*). In odor recognition, when met with failure, it is quite natural for subjects to rely on sensory clues which are not olfactory origin (*85,86*). Addition behavioral factors such as the usage of rewards, treats, and gifts to enhance subject's motivation, may play a role in the overall performance of participating subjects. Using beer as a carrier, Brown et al. (*87*) review three reference methods for detection threshold, and summarized that a panel size of 25 or more test subjects would be necessary for general validity of results; and training of test subjects improved reliability of results. Also, the authors did not find significant relationship between threshold degree of sensitivity and age, sex, smoking habits, and ethnic origin.

However, sensitivity to a specific substrate in solution, does not reflect improved perception of such substrate in the complex food and beverage systems. Beyond detection, recognition of substrates depends on physiological status, behavioral pattern, and cognitive level of test subjects (*84*). A relatively simple beverage such as tea, contains complex aroma, tactile, and taste components which require training and practices in order to describe its sensory qualities. It is apparent that a Chinese who eats rice on a daily basis will recognize the aroma, taste, and flavor of rice a great deal better and faster than a Swede who eats rice when he visits a Chinese restaurant maybe twice in his life time. Naturally, subjects with nasal congestion, will find difficulty in breathing, not to mention sniffing an odor. Recognition of aroma, taste, and flavor is a learned process, based on exposure and experience (*88-93*). In complex systems, threshold values provide little quantitative value. However, in clinical screening for hyposmia among the elderly and other susceptible population, it is a powerful method (*94*).

Paired comparison, duo-trio, and triangle tests are often termed difference tests, after their design. In paired-comparison, two samples are presented simultaneously, and subjects are asked to indicate which of the two elicit the greater intensity of sensory characteristic. Duo-trio involves three samples, but subjects are first presented with a labeled standard or reference sample, followed by two additional samples in randomized order, of which one of them is also a standard. Subjects are required to indicate which sample of the pair was the same as the standard. Triangle tests also use three samples, but unlike, duo-trio, the samples are presented simultaneously. Two of the three are identical, and subjects are required to indicate the odd sample. These tests require subjects to choose one sample, thus often results are forced-choice when the choice is not clear.

In the case of paired comparison, the probability of choosing the correct sample, akin to tossing of a coin, is 0.50. Duo-trio test is very similar to paired comparison in that subjects are actually testing one pair at a time, requiring evaluating two pairs of samples induce twice the amount of behavioral input, but the probability of a correct choice remain at 0.50. Triangle tests allow subjects to evaluate samples freely, and frequently, therefore the probability of choosing of the odd sample, is 0.333 (*95*).

The advantages of using paired comparison test include, simplicity in design, and sensitivity in determination of differences. It is particularly useful when the objective of study is not centered on determining the parameter of the difference between samples - a quick and dirty determination of difference. Additionally, these tests are powerful in measuring the perceptive abilities of test subjects to discern subtle differences. The limitation lies within the advantage where it often provides a minimum amount of information on sensory characteristics causing the difference. Duo-trio test is similar to paired comparison in its advantages, but is less sensitive because of the tendency for reversal of judgement. It is also slower, requiring two steps instead of one. Triangle tests provide better probability than its' counterparts, but similar to Duo-Trio, it is less sensitive and require a lengthier test session.

However, Tuorila et al. (*96*) reported that among the discrimination tests of duo-trio, paired comparison, constant stimulus and triangle tests, the later test was found to be most sensitive in subjects' detection of vanillin, butyl acetate, trimethyl amine, diacetyl, and limonene. In conducting difference tests, it is vital for researcher to adequately randomize the samples. Without proper randomization, subjects will be able to choose the "same" or "odd" sample from the order of presentation without evaluation of samples. Estimation of statistical significance is based on statistical tables (97), where the number of correct responds provided by test subjects, are matched against the numeric values in the tables.

One important application using pair comparison method is in determination of relative sensory stimulus, such as equal-sweetness between sucrose, fructose, glucose, aspartame, saccharin etc (98-102). In determination of relative sensory stimulus, the estimation of stimulus concentration equivalent can be carried out using the proportional method (53), or the regression method (103). Proportional method utilize a formula where the stimulus concentration is factored proportionally based on the number of subjects reporting greater sensory stimulus (sweeter) of test solution (glucose) over a given constant (sucrose). Larson-Powers' regression method uses the linear regression equation to determine the stimulus concentration where 50% of responses indicate test solution (glucose) to be sweeter than the given constant (sucrose). Determining relative equivalent in stimulus is an important factor in industries considering utilization of certain ingredient, for example, in establishing the concentration of saccharin, and more recently aspartame, that is as equally sweet as its nutritive counterpart in beverages.

Lastly, ad-libitum mixing method (27) is often used to determine the ability of test subjects to discriminate difference or determine the preferred intensity of stimulus. In a discriminatory capacity test subjects freely mix a low intensity (water) with a high intensity (10% sucrose solution) stimulus to equate the intensity of a given standard (3% sucrose solution). Similarly, in preference test, subjects would mix a low intensity with a high intensity stimulus to their preferred level. Ad-libitum method can be a powerful tool in screening of test subjects for their acuity in perception of certain taste, and their ability to reproduce responses. It is simple and straight forward to administer. Furthermore, it can be carried out without imposing a force-choice scenario, and best of all requires little training of subjects.

Quantitative Tests. Ranking, scaling, and magnitude estimations are routinely used in quantitative evaluation of sensory attributes. Ranking of samples in an increasing order perception intensity was introduced by Kramer (104). The author claimed that the use of non-parametric statistical analysis method was not restricted by the assumptions underlying the analysis of variance, is therefore more accurate for application to analysis of sensory data while retaining the power of analysis of variance. The advantage of using rank order method, lies in its simplicity, allowing large amount of data to be analyzed for statistical validity. It is suitable for use in situations where actual values are not meaningful, therefore, convenient to rank a series of samples in order of preference or difference. Because of errors found in Kramer's probability table, Joanes (105) recommended Friedman rank sum test be used in analysis of data obtained from ranking methods. However, Roessler (106) reported that analysis of variance may be a reliable choice in analyses of rank order data, but it lacks the ability to test variability between subjects.

Scaling methods utilize scales that can be unstructured or structured. Structured scales are straight forward to use; easily understood by test subjects when attributes are clearly defined; versatile; diverse in usage; and are reproducible. Unstructured scales provide all the above-mentioned application of structured scales, with the additional advantage of reducing number preference and bias; and elimination of difficulties of developing descriptive language.

Scales can be numerical or non-numerical. Numerical scales provide data that are easy to analyze and interpret, thus extremely useful in establishing the physical treatment that results in proportional changes in perception. However, care should be taken to prevent specific and/or round number bias; supply a meaningful amount of numbers; and establish a linear relationship between the numbers and the concentration. Non-numerical scales as stated earlier, avoids number bias and having to define descriptive terms. Computerization of data collection eliminates drawback of unstructured scales caused by the length of time required to convert data into numbers for statistical purpose.

Magnitude estimation is a type of single-directional ratio scaling, which measures the relationship between the physical and sensory continuum in a proportional manner. Test subjects estimate the magnitude of sensory attributes based on a given reference. For example, a sample that is three times brighter, rougher, larger, sweeter, etc., than the reference with numerical designation of 10, should receive a score of 30. Since perception is anchored to a reference, magnitude estimation scaling provides infinite numbers, proportional judgements, easy conversion between scales or to percentages, and reliability over category scaling. However, this method requires training and practice to keep test subjects from reverting to ranking, category scaling or a combination of category ratio scales, and number bias. Furthermore, various normalization procedures used in data analyses eliminates subject variability, and therefore could yield results with erroneous impressions of subjects agreement. It should not be used in hedonic responses, which are bi-directional (like-dislike).

Galanter (84) set four criteria upon which sensory scales should be accepted and compared the performance of category scale and magnitude estimation. The scale should provide consistent repeatability of results; be able to explain the result in terms of a basic theory; allow prediction of new findings based on the results; and provide invariance of result upon manipulation of ostensibly nonessential characteristics of the experiment. He used these criteria to compare category and magnitude scaling, and reported that there are no apparent difference between the two scaling methods in the first three criteria. However, closer examination showed that the two scaling techniques give different results. The category scale is distorted when the given stimuli used for construction are changed. On the other hand, magnitude scale is invariant with changes in stimulus ensemble. Thus the category scale value for a particular stimulus is not as intimately associated with the stimulus as is the magnitude value. This is how many psychophysicists come to believe that the magnitude scale reveals more about the sensory effects of stimuli, and therefore more about the bases of the judgmental process of people when they are called upon to act with respect to the magnitudes of stimuli in their environment (*84*).

In measurement of temporal qualities of sensory perception, time-intensity is the method of preference. It provides information on dynamics of sensory characteristics over time, which is independent of quality or overall intensity of stimuli. At its infancy, the time-intensity method was reserved for subjects endowed with precarious eye-hand-feet coordination, to draw out their responses on a moving chart recorder. With the onslaught of computers and speedy information processing systems, it has evolved into an effective research tool. Thanks to computers and info-technology, time-intensity method has been simplified to sliding of the "joy-stick", turning of the "knob", or gliding of the "mouse". Computerization also eliminated many problems associated with data transformation (*107,108*). Typically, time-intensity study provide data graphically from time of onset, maximum intensity, time to reach maximum, to total duration of sensation, as well as the rates of onset and decay of perception. Its ability to display two dimensional response, makes it an indispensable method in determining a total temporal "picture" of sensation perceived during mastication, oral manipulation, swallowing, oral physiological responses, change in physical and chemical properties of samples and many more. However, perception of changes in physical and chemical properties must occur within a fairly short duration - minutes.

Qualitative Tests. Characterization of "quality" attributes falls within the realm of qualitative tests or descriptive sensory analysis. By definition, descriptive analysis identifies and quantifies sensory attributes of products. It requires the development of descriptive adjectives or descriptors to characterize the qualitative aspects of appearance, odor, taste, and textural attributes associated with complex changes in products (83). In order to develop effective descriptors panelists are extensively trained to master appropriate sensory techniques and analytical methods, to detect, identify or qualify, and quantify sensory qualities of products under investigation.

There are several techniques of descriptive analysis designed for a wide range, or limited set of sensory attributes. ASTM recently published a short manual that consists of four descriptive methods (109). The best known method, the flavor profile method, developed by the Arthur D. Little Co., was used to test single sensory component of a substance (110). Aroma, taste, flavor, tactile factors, and aftertaste are rated for intensity on a four category and point scale of 0,)(, 1, 2, and 3. The Tilgner dilution profile utilizes identification of stimulus at threshold concentration, and expressed as percent dilution as a measure of odor or flavor of substances (111). The method assumes that diluted foods have flavor properties directly related to those of the undiluted product. Texture profile of General Foods, is designed for tactile attributes (112). Tragon's quantitative descriptive analysis (QDA) was developed for quantitative measures of flavor attributes (113-115). Deviation-from-reference profile method describes characteristics of sensory attributes systematically and objectively, without prejudice due to preference. A sensory attribute "vocabulary" is utilized that is applicable to different types of a specific product, e.g., beer in general. Recently, methods with varying technique in development of descriptive terms, such as the free-choice profiling has been reported to be useful, and surely many more variation of descriptive methods would be developed in the future.

The above mentioned techniques differ in their applications for testing varying type of sensory attribute (aroma, taste, texture, flavor), method of quantitative measure (rating, scoring, scaling), method of determining descriptive reference (group consensus, free-choice), experimental design (single or repeated measures), statistical method (group consensus, truncated mean, analysis of variance, multivariate analysis-principle component analysis), and data presentation (histogram, polar coordinates, Cartesian coordinates, multi-dimensional projection). Due to their variabilities in design and application, the choice of technique should be dependent upon whether the technique will be able to appropriately meet the researcher's objective.

The one commonality between all descriptive analyses is the utilization of very well trained test subjects. Subjects should be skilled in detection, identification, description, recognition, quantitative evaluation of sensory attributes in the product. Therefore, selection, training of subjects, and practice for accurate description of sensory attributes are essential (53). Schemper et al. (93) reported that age may be one of the factors which influence aroma identification. Retrieval, encoding, and spontaneous verbal mediation processes seem to be heavily compromised among the older subjects. However, sluggish identification can be improved with labeling of sample. Since descriptive terms are used in describing sensory attributes, it is essential that these terms be accurately defined to achieve uniformity in description among test subjects. The process of developing a qualitative sensory panel is both labor and time intensive. To sum it up in a sentence, quality evaluation can not be carried out in an uncontrolled environment, without clear definition of reference or standard descriptors, nor by untrained subjects.

Within the fields of viticulture, enology, and brewing, reports on characterization of flavor (aroma, taste, and tactile) qualities of grapes (17,116,117), wine (118-129), and beer (121,130,131) are well documented. Readers who are interested in conducting qualitative sensory evaluation are advised to consult sensory scientists with knowledge in such techniques. Where multivariate analysis is involved, statistical and computer support would be a requirement as well. In data presentation, it is a good practice to use a combination of methods of display, to enhance comprehension and clarity. It is important to note that qualitative analysis is only a part of sensory research. It is obvious this method should not be used as routine quality control (QC), as it would be akin to killing an ant with an elephant gun.

Akin to conducting research experiments, the choice of sensory method in routine QC work, should be within the framework of the product industry's QC objectives. For example, if the objective is to measure consistency of products within a set of established quality standards, deviation from standard would be quite effective. On the other hand, if measurement of specific sensory attribute (eg. a specific taste, proprietary aroma etc.) is required, more appropriately, quantitative methods should be employed. However, in development of new products, it may be necessary to identify the various positive or negative sensory attributes that are directly related to product quality. In such cases, it would be most appropriate to invest time and resource to establish a descriptive panel and conduct intricate profile work on the product.

"Some Don'ts"

Pangborn (*80*) conducted an extensive review of literature on sensory analyses and reported the most common "sins" committed in sensory science. They are: using consumer tests with analytical laboratory judges; using analytical tests with untrained consumers; misuse of difference testing; and inappropriate scaling procedures. Since sensory evaluation is based on utilizing human subjects as "instrument", it is clear that tests procedures should be set within the limitation of human capabilities.

The first type of misuse involves utilization of trained laboratory subjects to do consumer affective tests. In this case, the objective of training is defeated. After all the time and effort spent in training and practice, subjects are now asked to revert back to the level of understanding at time zero. This is much akin to using a High Potential Liquid Chromatographic instrument to weigh a sample.

The second type of misuse is expecting an untrained person to do objective evaluation of samples, without clear direction and training, much like asking people to define "quality".

Misuse of difference testing and inappropriate scaling procedures are common because of the array of methods in sensory evaluation. It is not an easy task for researchers not well acquainted with sensory evaluation to choose a method, to utilize appropriate procedure to carry out tests to meet the set objectives of experiment. Abuse of difference tests in conducting consumer acceptance test, frequently stems from assumption that a lack in statistical significance reflects a lack of difference between samples. Thus, if the products are preferred equally, they are indistinguishable. Stone et al., (*132*) reported that a common error in analyzing data from difference tests stems from improper application of probability tables. Abuse of scaling methods involves using non-linear scales. For example setting numbers to equate descriptions which are not related to each other in linear proportion, 1=flowery, 2 = skunky, 3 = fruity etc. When scales are divided into unequal proportion, the results also spells disaster. Other type of errors such as number bias, central tendency, insufficient points, etc., have been mentioned earlier.

Using magnitude estimation, which measures in one direction, to determine hedonic responses, which are bi-directional (like-dislike), and thus vary immensely from subject to subject, usually results in bimodal or multimodal distribution with varying intensity of sensory attributes.

Often quantitative results are wantonly interpreted to reflect the overall quality, and consumer preference of a product. Without prior qualification of sensory attributes measured, a sample that is perceived as brighter, sweeter, larger, etc., would have little to do with the "quality" and less with consumer preference.

Koster (*133*) provided numerous scenarios where researchers utilized sensory methods incorrectly to meet set test objectives, thus results obtained are often open to peer criticism. The author listed three addition type of "pitfalls" in sensory analysis. They are: over-estimation of human capabilities; neglecting the influence of context; and using insufficient selection procedure. The author stressed the importance of taking human limitation into account in designing of tests, i.e., sensory fatigue, motivation factor, physiological limitations etc. Care should be taken to account for context effect, such that the effect of placing subjects in test booths in conducting consumer acceptance tests, should not be ignored. Selection of effective subjects for conducting descriptive analysis can not be over emphasized.

"A Good Place to Start"

In closing, the following is a check list that covers some of the fundamental components in planing and conducting sensory evaluation. Of course it is not an absolute precept, nor a conclusive list of items to understand and/or control, rather for readers to use as a guideline towards a positive experience in sensory evaluation. There will always be additional considerations as individual studies varies in goal, objective, product form, and the researcher's approach to solving problems.

Check-list
• Are you aware of the "sins" and "pitfalls" of sensory evaluation?
• Have you conducted literature search and review?
• Do you have access to an expert opinion?
• Do you know your goal and hypothesis?
• Have you set the objectives of your study?
• What type of measurement are you making to meet your objectives?
• How many samples and treatments will be involved?
• Do you know which experimental design will fit your needs?
• Did you verified your experimental design with a statistician?
• Is the method designed or suitable, to measure your parameters of interest?
• Will you be carrying out chemical or instrumental analyses to correlate your sensory findings?
• Are you aware of the limitation(s)with your sensory method of choice?
• Do you know how to set test procedure in accordance with the method(s) chosen?
• Did you tried the test procedure?
• Is there an aftertaste problem?
• Is there fatigue problem?
• What are you doing about sensory fatigue?
• Is rinsing of mouth part of your test procedure?
• How long will each session last and how many sessions?
• Do you know the idiosyncracy of your product?
• Is it a food product (complex system) or a substrate stimulus (model system)?
• At what concentration will you be testing substrates?
• Is the substrate stable in the form (solution, carrier air etc.,) of delivery?
• Are samples received good representatives of the product?
• Are there sampling problems?
• Where did the samples come from?
• What kind of processing have the samples undergone?
• What type of processing will the sample undergo?
• Will processing cause variability within and between samples?
• Did you tried the samples?
• Have you bench•top samples?
• What sensory characteristics are of interest?
• Do you need to define these sensory characteristics?
• Do you need to visually mask samples?
• Did you provide enough sample for your subjects to evaluate?
• Will extensive training of subjects be required?

- How many subjects must finish the test for statistical validity?
- Did you provide sufficient training?
- What is the background (social, culture, ethnic etc.,) of your test subjects?
- How will you keep them motivated?
- Is there a communication problem?
- Did you check for repeatability of your subjects' responses?
- Is the test procedure fairly simple to follow?
- Are you sure your subjects are following your procedure?
- Will the tests be conducted in booths or controlled environment?
- Are there adequate controls on temperature and humidity of test site?
- Are there adequate lighting, ventilation, and sound proofing?
- How will you analyze your data within your experimental design?
- Do you have the capability or support to conduct your data analysis?
- How will you interpret your result?
- Is your interpretation within the context of your methodology and experimental design?
- Did you meet your objectives, and goal?
- Is your hypothesis tested?
- Do you think you did a good job?

Acknowledgments

In remembrance of RMP.

Literature Cited

1. ASTM STP #434, *Am. Soc. Test. Mat.*, Philadelphia, 1968.
2. Blair, J.R. *Food Technol.* **1978**, 32(11), 61-62.
3. Pearce J. *Food Technol.* **1980**, 34(11), 60-62.
4. McDermott, B.J. *Food Technol.* **1991**. 44(11), 154, 156, 158.
5. *Food Acceptance and Nutrition. A series of monographs;* Solms, J.; Booth, D.A.; Pangborn, R.M.; Raunhardt, O. Eds. Academic Press: New York, 1987; 29-189.
6. Rozin, P. *Nutrition Review,* **1991**, 48(2), 107-113.
7. Rolls, B.J.; Rolls, E.T.; Rowe, E.A.; Sweeney, K. *Physiol. and Behavior,* **1981**, 27,137-142.
8. Booth, D.A.; Lee, M.; McAleavey, C. *Br. J. Psychol.* **1976**, 67, 137.
9. Shuichi, K. *Nutr. Rev.* **1992**, 50(12), 421-431.
10. Bedichek, R. In *The Sense of Smell*; Doubleday, New York, 1960.
11. Catheart, W.H.; Killen, E. *J. Food Research,* **1940**, 5, 307.
12. Amerine, M.A.; Roessler, E.; Fillipello, F. *Hilgardia.* **1959**, 28,477-556.
13. Ough, C.S.; Baker, B.A. *Hilgardia,* **1961**, 30, 587-619.
14. Ough, C.S.; Winton, W.A. *Am. J. Enol. and Vit.* **1976**, 27, 136-144.
15. Tromp, A.; Conradie, W.J. *Am. J. Enol. and Vit.* **1979**, 30, 278-283.
16. Wu, L.A.; Bargmann, R.E.; Powers, J.J. *J. Food Science.* **1977**, 42, 944-952.
17. Noble, A.C. *Am Wine Soc Journal,* **1984**, 16(2), 36-37, 57.
18. Noble, A.C. *J Science Food Agri.* **1990**, 53(3), 343-353.
19. Pangborn, R.M. *Food Technol.* **1964**, 18(9), 63-67.
20. Dawson, E.H.; Harris, B.L. *U.S. Dept. Agr., Agr. Inf. Bull. No. 34.* **1951**, 134.
21. Dawson, E.H.; Brogdon, J.L.; McManus, S. *Food Technol.* **1963a**, 17(9), 45-48.
22. Dawson, E.H.; Brogdon, J.L.; McManus, S. *Food Technol.* **1963b**, 17(10), 39-41,43-44..
23. Noble, A.C. *Food Technol.* **1975**, 29(12), 56-60.
24. ASTM STP #440, *Am. Soc. Test. Mat.*, Philadelphia, 1968.
25. Szczesniak, A.S. *J. Food Science,* **1963**, 28, 385-389.

26. Gatchalian, M.M. In *Sensory Evaluation Methods with Statistical Analysis*; University of the Philippines: Diliman, Quezon City, Philippine, 1981.
27. Pangborn, R.M. In *Modern Methods of Food Analysis;* Steward, K.K.; Whitaker, J.R. Eds.; AVI: Westport, Conn. 1984, 2-28.
28. Stevens, M.A. *HortScience,* **1970**, 5, 95.
29. Spencer, M.D.; Pangborn, R.M.; Jennings, W.G. *J. Agr. Food Chem.* **1978**, 26(3), 725-732.
30. Stevens, M.A.; Albright, M. *HortScience.* **1980**,15,48-50.
31. Wood, M. *Agr. Research,* **1992**, 40(2), 22.
32. Sanchez, N.B.; Lederer, C.L.; Nickerson, G.B.; Libbey, L.M.; McDaniel, M.R. In Food Science and Human Nutrition; Charalambous, G. Ed., Elsevier: New York, 1992, 371-402.
33. Dethmers, A.E. *Food Technol.* **1979**. 33(9), 40-42.
34. Skinner, E. *Food Technol.* **1980**, 34(11), 65-66.
35. Moskowitz, H.R.   In *Product Testing and Sensory Evaluation of Foods, Marketing and R & D Approaches;* Food & Nutrition Press: Westport, Conn, 1983.
36. Carter, K.; Riskey, D. *Food Technol.* **1990**, 44(11), 160, 162.
37. Cohen, J.C. *Food Technol.* **1991**, 44(11), 164,166,174.
38. Stone, H.; McDermott, B.J.; Sidel, J.L. *Food Technol.* **1991**, 45(6), 88,90,92-95.
39. Ellis, B.H. *Food Product Development,* **1970a**, 4(5), 86.
40. Ellis, B.H. *Food Product Development,* **1970b**, 4(6), 46.
41. Hoskins, W.A. *Food Technol.* **1971**, 25(4), 99-101.
42. Civille, C.V. Food Technol. **1978**, 32(11), 50-60.
43. Erhardt, J.P. *Food Technol.* **1978**, 32(11), 57-58, 66.
44. Tassan, C.G. *Food Technol.* **1980**, 34(11), 57-59.
45. Wren, J.J. In *Sensory Quality Control, Practical Approaches in Food and Drink;* The Food Group and the Society of Chem. Ind, London, 1977.
46. Gatchalian, M.M.; De Leon, S.Y. In *Introduction to Food Technology, Emphasis on Production and Quality Control. Vol. I*; University of the Philippines, Diliman, Quezon City, Philippines, 1975, 1-18.
47. Kramer, A.; Twigg, B.A. *Quality Control in the Food Industry;* AVI: Westport, Conn. 1973; Vol I, 253-283.
48. Kramer, A.; Twigg, B.A. *Quality Control in the Food Industry;* AVI: Westport, Conn. 1973; Vol II, 157.
49. Zelek Jr., E.F. *Food Technol.* **1991**, 44(11), 168,170,172,174.
50. Prell, P.J. *Food Technol.* **1976**, 30(11), 40, 42-44, 46.
51. Kramer, A. In *Handbook of Food and Agriculture*; VanNostrand Reinhold: New York, 1955, 733-750.
52. Muller, H.G. *J. Texture Stud.* **1969**, 1, 38-42.
53. Amerine, M.; Pangborn, R.M.; Roessler, E.B. In *Principles of Sensory Evaluation of Food;* Food Sci. & Tech. Monograph; Academic Press, New York, 1965.
54. Konosu, S.; Hayashi, T; Yamaguchi, K. In *Umami, A Basic Taste*; Kawamura, Y; Kare, M.R. Eds.; Marcel Dekker, New York, 1987, 235-269.
55. *Sensory Analysis of Foods*; Piggott, J.R. Ed; Elsevier's: New York, 1988; Vol I.
56. Meilgaard, M.D.; Civille, G.V.; Carr, B.T. In *Sensory Evaluation Techniques;* 2nd ed; CRC Press: Boca Raton, Fl, 1991.
57. Clydesdale, R.M.; Francis, F. *J. Food Product Development,* **1968**, Aug.-Sept., 30.
58. Meiselman, H.L. In *CRC Critical Reviews in Food;* Furia, T.E. Ed.; CRC Press: Boca Raton, Fl, 1972.
59. Le Magnen, J.; MacLeod, P. *Olfaction and Taste VI*; Information Retrieval Ltd, London, 1977.
60. Hornstein, I.; Teranishi, R. *Chem. and Eng. News,* **1967**, 45, 93-108.

61. Trinchese, T. *Food Product Development*, **1968**, 2(2), 72-74.
62. Larmond, E. *Food Technol.* **1973**, 27(11), 28, 30, 32.
63. ASTM STP #E480, *Am. Soc. Test. Mat.*, Philadelphia, 1973.
64. Girardot, N.F.; Peryam, D.R.; Shapiro, R. *Food Technol.* **1952**, 6(4), 140.
65. Pangborn, R.M.; Dunkley, W.L. *Dairy Science Abs.* **1964**, 26(2), 55-62.
66. Amerine, M.; Roessler, E.B. In*Wines, Their Sensory Evaluation*; Freeman: San Francisco, 1976, 116-121.
67. ASTM STP #758, *Am. Soc. Test. Mat.*, Philadelphia, 1981.
68. Sacharow, S.; Griffin, R.C. In *Principles of Food Packaging. 2nd ed.*; AVI: Westport, Cn, 1980.
69. Scoville, W.L. *J. Am. Pharm. Assoc.* **1912**, 1, 453.
70. Gillette, M.H.; Appel, C.E.; Lego, M.C. *J. Food Science.* **1984**, 49, 1028-1033.
71. Govindarajan, V.S.; Narasimhan, S.; Khanaraj, S. *J. Food Science and Technol.* **1977**, 14, 28.
72. Nasrawi, C.W.; Pangborn, R.M. *J. Sensory Science*, **1989**, 3, 287-290.
73. Cliff, M.; Heymann, H. *AChemS.* abstract, **1992**, 17, 605.
74. Nasrawi, C.W.; Pangborn, R.M. *Physiol. and Behavior*, **1990**, 47, 616-623.
75. Sidel, J.L., Stone, H. *Food Technol.* **1976**, 30(11), 32-38.
76. Little, T.M.; Hills, F.J. In *Statistical Methods in Agricultural Research*; Wiley: New York, 1980.
77. Bender, F.G.; Douglass, L.W.; Kramer, A. In *Statistical Methods for Food and Agriculture;* AVI: Westport, Cn, 1982.
78. Sidel, J.L.; Stone, H.; Bloomquist, J. In *Sensory Quality in Foods and Beverages, Its Definitions, Measurement and Control;* Williams, A.A.; Atkins, R.K. Eds., Ellis Horwood Ltd: Chichester, England, 1983.
79. Stone, H.; Sidel, J.L. In *Sensory Evaluation Practices;* Academic Press: Orlando, FL, 1985.
80. Pangborn, R.M. *Cereal Foods World*, **1980**, 25(10), 637-640.
81. Elrod, J. *Food Technol.* **1978**, 32(11), 63.
82. ASTM STP #682, *Am. Soc. Test. Mat.*, Philadelphia, 1979.
83. Pangborn, R.M. In *Principles of Sensory Analysis of Food, Food Science and Technology 107 Lecture Syllabus*; University of California, Davis, California, 1986.
84. Galanter E. In *New Direction in Psychology*; Holt, Rinehart & Winston, NY, 1962, 89-156.
85. Hall, R.L. In *Flavor Research and Food Acceptance;* Little A.D. Ed.; Reinhold, New York, 1958, 224-240.
86. DuBose, CN.; Cardello, A.V.; Maller, O. *J. Food Sci.* **1980**, 45,1394-1399.
87. Brown, D.G.W.; Clapperton, J.F.; Meilgaard, M.C.; Moll, M. *Am. Soc. Brew. Chem.* **1978**, 36, 73-80.
88. McAuliffe, W.K.; Meiselman, H.L. *Perception and Psychophysics,* **1974**, 16, 242-244.
89. O'Mahony, M.; Goldenberg, M.; Stedman, J.; Alford, J. *Chemical Senses and Flavor,* **1979**, 4, 301-318.
90. Cain, W.S. *Science,* **1979**, 203, 276-281.
91. Cain, W.S. *Psychology Today,* **1981a,** July, 48-55.
92. Cain, W.S. *Chemical Senses,* **1981b**, 7(2), 129-142.
93. Schemper, T.; Voss, S.; Cain, W.S. *J. Gerontology,* **1981**, 36, 446-452.
94. Amoore, J.E. *Chemical Senses and Flavors,* **1977**, 2, 267-281.
95. Pfaffmann, C. In *Food Acceptance Testing Methodology;* Peryam, D.R.; Pilgrim, F.J.; Peterson, M.S. Eds.; NAS-NRC: Chicago, IL, 1954.
96. Tuorila, H.; Kurkela, R.; Siuhdo, M.; Suhonen, U. *Lebensmittel-Wissenschaft-und-Technologie,* **1982**, 15, 97-101.
97. Roessler, E.B.; Pangborn, R.M.; Sidel, J.L.; Stone, H.J. *J. Food Science,* **1978**, 43(3), 930-943.
98. Cameron, A.T. In *The Taste Sense and the Relative Sweetness of Sugar and other*

Sweet Substances; Sci. Rep. Ser. #9, Sugar Research Foundation: New York.
1947.
99. Pangborn, R.M.; Gee, S.C. *Nature.* **1961**, 191(4790)810-811.
100. Pangborn, R.M. *J. Food Sci.* **1963**, 28(6)726-733.
101. Stone, H.; Oliver, S. *J. Food Sci.* **1969**, 34(2),215-222.
102. Yamaguchi, S.; Yoshikawa, S.; Ikeda, S.; Ninomiya, T. *Ag. & Biol. Chem.*
1970, 34(2), 181-186.
103. Larson-Powers, N.; Pangborn, R.M. *J. Food Sci.* **1979**, 43(1),41-46.
104. Kramer, A.; Kahan, G.; Cooper, A.; Papavasilian, A. *Chem Senses & Flavor*,
1974, 1,123-133.
105. Joanes, D.N. *J. Food Science*, **1985**, 50, 1442-1444.
106. Roessler, E.B. In *Principles of Sensory Analysis of Food, FS&T 107 Laboratory
Manual*; Pangborn, R.M. Ed., **1986**, University of California, Davis, 97-104.
107. Schmitt, D.J.; Thompson, L.J.; Malek, D.M.; Munroe, J.H. *J. Food Science.*
1984, 49, 539-542, 580.
108. Guinard, J.X.; Pangborn, R.M.; Shoemaker, C.F. *J. Food Science.* **1985**, 50,
542-544.
109. *Manual on Descriptive Analysis of Testing for Sensory Evaluation.* Hootman,
R.C. Ed., ASTM, Philadelphia, PA. **1992**.
110. Caul, J.F. *Advances in Food Research*, **1957**, 7, 1-40.
111. Tilgner, D.J. *Food Technol.* **1962**, 16(2), 26-29.
112. Civille, C.V.; Szczesniak, A.S. *J. Texture Studies*, **1973**, 4, 204-223.
113. Sidel, J.L.; Woolwey, A.; Stone, H. In Fabricated Foods, Inglett, G.E. Ed.;
AVI: Westport, CT., 1975.
114. Stone, H.; Sidel, J.; Oliver, S.; Woolsey, A.; Singleton, R. *Food Technol.*
1974, 28(11), 24, 26, 28, 29, 32, 34.
115. Stone, H.; Sidel, J.L.; Bloomquist, J. *Cereal Foods World*, **1980**, 25, 10, 642-
644.
116. Carroll, D.E. In *Evaluation of Quality of Fruits and Vegetables*; Pattee, H.E. Ed.;
AVI: Westport, CT., 1985, 177-197.
117. Morris, J.R. In *Evaluation of Quality of Fruits and Vegetables*; Pattee, H.R. Ed.;
AVI: Westport, CT., 1985, 129-176.
118. Amerine, M.A. *Food Technol.* **1982**, 110, 106-108.
119. Goniak, O.J.; Noble, A.C. *Am J Enol Vit*; **1987**, 38(3), 223-227.
120. MeCredy, J.; Sonnemann, J.; Lehmann, S. *Food Technol.* **1974**, 28(11), 36-41.
121. Meilgaard, M.C.; Dalgliesh, C.E.; Clapperton, J.F. *J. Am. Soc. Brew. Chem.*
1979, 37, 47-52.
122. Noble, A.C. *Vini d'Italia*, **1981**, 23(135), 325-340.
123. Noble, A.C.; Williams, A.A.; Langron, S.P. In *Sensory Quality in Foods and
Beverages, Definition, Measurement and Control;* Williams, A.A.; Atkin, R.K.
Eds; Ellis Horwood Ltd: Chichester, 1983.
124. Noble, A.C.; Williams, A.A.; Langron, S.P. *J. Science Food Agr.* **1984a**, 88-
98.
125. Noble, A.C.; Arnold, R.A.; Masuda, B.M.; Pecore, S.D.; Schmidt, J.O.; Stern,
P.M. *Am. J. Enol. and Vit.* **1984b**, 35, 107-109.
126. Noble, A.C.; Arnold, R.A.; Buechsenstein, J.; Leach, E.J.; Schmidt, J.O.; Stern,
P.M. *Am. J. Enol. and Vit.* **1987**, 38, 143-146.
127. Aiken, J.W.; Noble, A.C. *Am. J. Enol. and Vit.* **1984**, 35, 196-199.
128. Singleton, B.L.; Zaya, J.; Trousdale, E. *Am. J. Enol. Vit.* **1980**, 31,14-20.
129. Schmidt, J.O.; Noble, A.C. *Am. J. Enol. and Vit.* **1984**, 34, 136-138.
130. Meilgaard, M.C.; Reid, C.S.; Wyborshi, K.A. *J. Am. Soc. Brew. Chem.* **1983**,
40, 119-128.
131. Clapperton, J.F. *J. Inst. Brew.* **1973**, 79,495-506.
132. Stone, H.; Sidel, J.L. *J. Food Science*, **1978**, 43, 1028-1029.
133. Koster, E.P. In *Criteria of Food Acceptance*; Solms, J.; Hall, R.L. Eds.; Foster
Verlag: Zürich, 1981, 240-252.

RECEIVED May 7, 1993

Chapter 4

Flavan-3-ols and Their Polymers
Analytical Techniques and Sensory Considerations

J. H. Thorngate, III[1]

Department of Viticulture and Enology, University of California, Davis, CA 95616

Bitterness is one of the predominant tastes perceived in wine; astringency also plays a major role in wine oral sensation. Those compounds responsible for bitterness and astringency are primarily phenolic in nature with the flavan-3-ols, (+)-catechin and (-)-epicatechin and their polymers, the procyanidins, being of greatest importance. This chapter will review the sensory research which has been done to characterize the flavan-3-ols' role in eliciting bitterness and astringency, and the chromatographic techniques used to separate and isolate these compounds.

With few exceptions aroma contributes in a much greater fashion to our overall sensory impressions of food products than does taste (*1*); however, this is not to downplay the importance of gustation. As has been often noted, sweet taste often serves as a marker for caloric sources whereas bitter taste provides for aversive ingestion behavior of toxic materials (*2, 3*; though cf. *4*). Of the so-called four "basic tastes" only sweetness, sourness and bitterness are of practical importance to wine. In addition to the taste sensations there exist other non-gustatory oral sensations (e.g.–temperature, viscosity) of which astringency is the most important to wine.

The Sense of Taste

The surface of the tongue is the major locus of taste transduction in man, although the mucosal surfaces of the esophagus, trachea and oral cavity also contain taste receptor cells (*5*). Three different types of mucosal protuberances called papillae, found on the tongue, are the location of the taste buds which contain the receptor

[1]Current address: Food Research Center, University of Idaho, Moscow, ID 83843

0097–6156/93/0536–0051$06.00/0
© 1993 American Chemical Society

cells. The fungiform papillae are located on the dorsal anterior two-thirds of the tongue while the (circum)vallate form the *linae terminalis*, delineating the anterior two-thirds from the posterior third (*6*).   The foliate papillae are found on the lateral surface of the tongue, although their presence in man is most likely vestigial (*6*) if they are indeed present at all (*7*).

The receptor cells of the fungiform papillae are innervated by the chorda tympani branch of cranial nerve VII, the facial, whereas the receptor cells of the vallate papillae are subserved by the ninth cranial nerve, the glossapharyngeal. Cranial nerve X, the vagus, primarily innervates the palatal and laryngeal regions. However, none of these nerves are exclusively gustatory afferants; the X cranial nerve response in particular appears to be confounded with mechanical stimulation (*8*).

Although it appears that the anterior portion of the tongue (especially the tip) is more sensitive to sweet taste with bitter taste having lower thresholds in the vallate taste buds (*9*), more recent investigations have revealed that the picture is not all that clear. Collings (*10*) found that the lowest threshold for bitterness was actually in the receptor cells of the soft palate. Certainly it has been known for a long time that individual papillae can respond to more than one "primary" taste (*11*).   The ability of the taste receptor cells themselves to respond to more than one gustatory stimulus has important implications for taste theory (*3*).

Currently the two prevailing schools of taste encoding theory can be defined as those who believe that there are four basic, or primary tastes (sweet, sour, salty and bitter) versus those who subscribe to a taste continuum theory. Many arguments have been presented by each side (*12-17*) and as Faurion and Vayssettes-Courchay (*18*) note, often with the same data. It is clear that this issue has yet to be fully resolved (*19, 20*).

Regardless of the system of encoding the taste response, it is now commonly accepted that the process of transduction involves either binding of the taste stimulus (for sweet or bitter) to a membrane-bound receptor on the apical portion of the taste receptor cell, or (for sour and salty) direct influence of the cation on the receptor cell, either by the blocking of ion-channels (in the case of H^+) or by direct influx (in the case of Na^+) (*21*).

Bitter taste is especially problematic; as Shallenberger and Acree (*22*) noted, "only limited correlations between bitter taste and molecular structure can be found." The compounds which can elicit bitterness include a diverse range of structures including such compounds as strychnine and caffeine (alkaloids); amygdalin (a glycoside from bitter almond); (-)-humulone (from hops); picric acid and urea.

Not only are the stimuli structurally diverse, the transduction mechanisms seem to be equally varied. While one pathway utilizes a membrane-bound receptor and G-protein cascade, other pathways appear to be receptor-independent and involve direct blockage of potassium channels (*23*). Kurihara *et al.* (*24*) have postulated that the non-ionic bitter sapophores are capable of adsorbing directly into and thus depolarizing the lipid membrane. That there are indeed multiple receptors for bitterness has been demonstrated with the thiocarbamides. Many persons are "taste-blind" to phenylthiocarbamide, while others find it intensely bitter; however, persons in both groups are sensitive to other bitter compounds.

Astringency

Sensory astringency (from the Latin *ad* (to) and *stringere* (bind)) is that oral sensation defined as a drying, or puckering, sensation, though not to be confused with a true taste sensation such as sourness.   As a tactile sensation the putative innervation is the V cranial nerve, the trigeminal, although the X cranial nerve might also respond to the drying of the mucosa (25).   As the trigeminal subserves the whole oral cavity the locus of sensation of astringency is considerably more diffuse than the lingual taste sensations.

The tongue and oral cavity are covered with a mucous membrane which is continually wetted by salivary secretions (26).   The sensation of astringency is thought to arise through the binding of the astringent principles to salivary glycoproteins and mucopolysaccharides with their subsequent precipitation. However, there might also be binding to structural proteins on the salivary ducts thus causing their constriction (27).   In either case the result is the aforementioned drying of the tongue and palate.

Sweetness, Sourness and Nonflavonoid Bitterness in Wines

In *Vitis vinifera* wines the major contributors to sweetness are glucose and fructose (although sucrose, galactose, ribose xylose, mannose and other sugars may be present in low concentrations) (28).   Glycerol invokes sweet taste at concentrations greater than 0.5%, and may be present in wines at concentrations ranging from 0.2-2.0%.   However, at these concentrations glycerol contributes little to the viscosity (29).   Amerine and Roessler (30) report ethanol as possessing a slight sweetness, and certain amino acids may also contribute to sweet taste.

The greatest contributor to acidity in wines is tartaric acid.   Tartaric acid and its acid salts are responsible for half or more of the normal acidity (28).   Malic acid is the other primary acid in wine, although citric, lactic, succinic and other organic acids in far lesser amounts are also present.   Noble *et al.* (31) found that, at constant pH and equal titratable acidity that lactic acid was perceived as more sour than citric, fumaric and tartaric acids; succinic acid was more sour than malic acid; and citric acid was less sour than fumaric, tartaric, malic and lactic acids.

Nonflavonoid bitterness in wine may result from a variety of sources, although none of them are of major significance.   The most important of these bitter sapophores would be tyrosol, which is produced by yeast fermentation.   It results from the deamination and decarboxylation of tyrosine, and is the only phenolic compound produced in significant amounts from nonphenolic precursors (32). While the tyrosol content of wine can reach 45 mg/L (28) the average tyrosol content is *ca.* 25 mg/L (33).   Aging has no effect on the tyrosol content.   The detection threshold in wine has not been determined, but in beer the threshold ranges from 10-200 mg/L, indicating that tyrosol may be contributing to the bitterness in wine (32). Dipeptides, long known to be sweet (34) also possess bitter analogs (35); however there has been no research to date on their role as bitter sapophores in wines.

As for the cinnamic and benzoic acid derivatives present in wine, the individual hydroxybenzoic acids are present in low levels (on the order of ppm) and thus have limited taste impact, although collectively they may contribute to wine bitterness (32). Vérette et al. (36) found that there was no taste contribution of the hydroxycinnamates in white wines when added at representative concentrations.

Crespo-Riera (37) noted that wines made from aromatic grape varieties were often more bitter than could be accounted for by the phenolic content, and proposed that the terpene glycosides were responsible. However Noble et al. (38) found that the terpene glycosides did not contribute to bitterness in Muscat wines when added back at representative levels.

Monomeric and Polymeric Flavan-3-ols in Grapes and Wines

The flavan-3-ols in wines are primarily (+)-catechin [2R, 3S] and (-)-epicatechin [2R, 3R]. (-)-Epicatechin-3-gallate is present in appreciable quantities in unripe seeds, but disappears over the course of ripening (39). The remaining flavan-3-ols present are the respective gallocatechin analogs of the above compounds (40).

The normal flavan-3-ol concentration in white wines ranges from 10-50 mg/L while in red wines they may reach 800 mg/L (33). Singleton (41) has estimated that for typical young white wines the flavan-3-ol content averages ca. 25 mg/L, while for young reds the content averages ca. 75 mg/L, although these values are, of necessity, rough estimates at best. These compounds are restricted to skins, seeds and vascular tissues, as opposed to the nonflavonoids which may also be found in the juice vacuoles (42). Romeyer et al. (43) found that (+)-catechin and (-)-epicatechin accumulated in the seed to a maximum concentration around veraison then diminished as the dimeric procyanidin content concomitantly increased.

The polymeric forms of the flavan-3-ols (more familiarly referred to as the procyanidins or condensed tannins) comprise the bulk of the phenolic material in red wines (32). In young wines the procyanidins appear to be primarily in dimeric and trimeric form, whereas in aged wines the relative degree of polymerization increases to 8-10 (44). Though polymers with molecular weights up to 7000 may still be soluble (45) the procyanidins in grapes most likely do not exceed a molecular weight of 3000 (41).

The polymeric flavan-3-ols most commonly have a 4-beta-8 or 4-beta-6 interflavan linkage (46), although other linkages are also known to occur in nature, such as the double linkage 4-beta-8; 2-beta-O-7 (47, 48). While in some plants either (+)-catechin or (-)-epicatechin predominate (sometimes almost exclusively), in grapes both monomeric forms are present in roughly equal amounts. Although there are three chiral centers in the heterocyclic ring of 4-linked polymers, the 2 position almost exclusively has the R absolute stereochemistry in the plant kingdom (49), and the 4 position always seems to be trans to the hydroxyl group at position 3. This still allows for isomerism about the 3 position, as well as the positional isomerism resulting from the linkage pattern. As acylation of the 3 position hydroxyl is also possible (typically with gallic acid) it is no wonder that Ricardo da Silva *et al.* (*50*) have recently isolated and identified 20 dimers and trimers from grape seeds.

Even so the chiral babel is not hopeless; it should be noted that in grapes (-)-epicatechin commonly serves as the chain extender with (+)-catechin serving as the terminal unit (*41*).   Furthermore, although 4-6 linkages occur roughly in the proportion 1:3 to 4-8 linkages (*51*), excessive branching due to 4-6 linkages do not seem to occur (*49*).   However, the interflavan bond is extremely susceptible to both acid and alkaline attack.   Therefore it is not clear as to whether or not all of the compounds identified are actually present in the seed, or whether some are experimental artifacts (*52, 53*).

It is clear that in grapes the bulk of the polymeric flavan-3-ols is found in the seeds, and to a lesser extent the stems, leaves and skins.   Values calculated from Kantz and Singleton (*54*) across four different cultivars indicate that 58.5% of the polymeric fraction was located in the seeds, 21% in the stems, 16.5% in the leaves and 4% in the skins, although there was much variability among the cultivars.

Under normal wine making practices neither the stems or leaves are in contact with the must thus the sole sources of procyanidins derive from the seeds and skins. Of these the skins are of greater practical importance as they are more readily extracted (*55*), though as the maceration time increases (as with red wines) the seeds play an increasingly important role as a procyanidin source (*56*).   Recent work indicates that the majority of the seed procyanidins is localized to the outer seed coat–the endosperm contains very little polymeric material (Singleton, V. L.; Thorngate, J. H., III, University of California, Davis, unpublished data).   It is important to note that the total procyanidin pool available, however, is reduced by incomplete extraction, adsorption or precipitation with solids and protein, conversion to nonphenolic products (i.e.–through oxidation), or polymerization to insoluble compounds (*57*).

The procyanidin content of wine turns out, not surprisingly, to be highly cultivar-dependent.   The white varietals have exceedingly low procyanidin contents (averaging < 10 mg/L GAE) whereas the red varietals averaged *ca.* 270 mg/L GAE of polymeric flavan-3-ols (*58*).   This has led to the recent realization that the anthocyanin sugar moiety serves to solubilize the procyanidins in anthocyanin-tannin polymers and thus keep them in solution (*58, 59*).

Monomeric and Polymeric Flavan-3-ol Sensory Properties

Rossi and Singleton (*60*) estimated the absolute threshold for monomeric flavan-3-ols in water to be 20 mg/L.   The taste threshold for (+)-catechin in deionized water was determined by Delcour *et al.* (*61*); they found that the best-estimate group threshold was 46.1 mg/L.   However, it should be noted that in both studies a detection threshold was determined and thus the perceptual nature of the stimulus is unknown, although Rossi and Singleton (*60*) noted that the primary taste perception from the flavan-3-ol fraction was bitterness.   Arnold *et al.* (*62*), however, found that in their study of phenolics in wine that all the fractions, including the monomeric, were both bitter and astringent.   This was also shown by Leach (*63*) in which (+)-catechin was added to white wine at increasing concentrations. Both bitterness and astringency increased with increasing (+)-catechin concentration; however the rate of increase of bitterness was significantly greater than that for astringency (64).

The monomeric flavan-3-ols are not technically chemical astringents as their molecular weight is less than 500 and they will not spontaneously precipitate proteins (57). However, whether or not they are sensory astringents remains unclear. Though sensory tests on phenolic fractions performed by Singleton et al. (60) have shown them to be bitter-only compounds, more recent studies have demonstrated a duality of oral sensation (62, 64). Whether or not this duality actually exists, or is a confounding artifact due to ethanol taste and oral sensation is the focus of ongoing research (Thorngate, J. H., III, University of California, Davis, unpublished data).

The procyanidins, however, have indeed been shown to be both bitter and astringent (62, 64, 65). Arnold et al. (66) found that on a per weight basis bitterness intensity increased with increasing degree of polymerization, as did relative astringency. The polymeric fraction (pentamers and larger) was calculated as being ca. 25 times more bitter than the monomeric fraction, and ca. 6 times more bitter than the trimer/tetramer fraction. Lea and Arnold (65) found an intensity maximum for bitterness in ciders with the (-)-epicatechin tetramer. However, it is important to note that the latter study did not standardize response on a per weight basis, and the relative concentrations of the fractions used could be confounding the responses.

The results are consistent with the theorized mechanisms of action for bitterness and astringency. The number of possible hydrogen-bonding sites increases with the degree of polymerization and thus the relative astringency should also be expected to increase. For bitterness either access to a putative membrane-bound receptor is being limited by the increase in molecular size, or else the difference in lipid-solubility of the trimers and tetramers allows them to depolarize the taste receptor cells directly, increasing their apparent bitterness (67). The bitter maximum noted by Lea and Arnold (66) may be artifactual, however, resulting from the astringency of the polymers masking their bitterness (though cf. 64).

Unfortunately there have been few temporal studies for bitterness and astringency exclusively using the flavan-3-ols. Robichaud and Noble (64) used time-intensity techniques (68) to study both the bitter and astringent responses to (+)-catechin and the procyanidins, whereas Fischer (69) studied the interactions of ethanol, pH and (+)-catechin concentration. Fischer's studies of astringency, however, utilized tannic acid, a hydrolyzable tannin (typically glucose fully esterified with gallic acid) not found in grapes, though its bitter and astringency properties appear similar to that of polymeric flavan-3-ols (64). Robichaud and Noble (64) found that the maximum intensity ratings for both bitterness and astringency increased with increasing concentration, as did the total duration of aftertaste. There was no change in the times to maximum intensity, however. Fischer (69) found that ethanol increased the time to maximum, the maximum intensity and the total duration of aftertaste for bitterness intensity of (+)-catechin and tannic acid.

A major and unresolved difficulty in assessing the sensory properties of the procyanidins lies in the fact that there is a cumulative carry-over effect for both astringency and bitterness (*70*). This has been elegantly demonstrated for astringency using time-intensity methodology (*71*). The order effect may be compensated by using reversed-order designs (*67*) or completely randomized designs with subsequent analysis of variance. Certainly designs which minimize the number of samples presented and include adequate rinsing regimens are to be preferred, and may serve to minimize the carry-over effect.

The continued use of procyanidins has important implications for the psychophysical studies of bitterness and astringency. Since the degree of polymerization provides a mock-continuum not otherwise available in gustatory stimuli the procyanidins could prove invaluable as sensory probes. Information obtained from their use could lead to characterization of at least one putative bitter receptor site and to an improved understanding of the mechanisms underlying astringency.

Methods of Analysis

Ribéreau-Gayon (*44*) succinctly summarized the general approaches required for the determining the nature of the procyanidins present: 1) chromatographic fractionation, 2) selective extraction, and 3) molecular weight determination. The number of articles discussing the analytical techniques for the separation and structural elucidation of the polymeric flavan-3-ols are legion and have been the subjects of several recent comprehensive reviews (*72-76*). The intent of this section, therefore, is to review those techniques which have been most widely used for the study of grape and wine procyanidins.

Although one-dimensional paper chromatography is of little utility for the separation of the procyanidins (*44*), two-dimensional techniques utilizing different separation mechanisms (partition vs. absorption being the most common) have proven quite successful. The dried chromatograms are sprayed with a 1:1 mixture of 1% potassium ferricyanide and 1% ferric chloride in 0.1 N HCl to produce Prussian blue spots indicative of oxidized phenolic compounds (*41*). Singleton *et al.* (*77*) used two-dimensional paper chromatography (butanol-acetic acid-water vs. aqueous acetic acid) to prepare reproducible maps of the phenolic components of grape seeds. The flavan-3-ol monomers (+)-catechin, (-)-epicatechin and epicatechin-3-gallate were positively identified; the oligomeric and polymeric forms were also in evidence, although at that time not identified as such. Paper chromatographic techniques have continued to be of qualitative importance, and are frequently employed for the identification of the monomeric through trimeric compounds (*72*).

The difficulty in quantification inherent to paper chromatography also extends to its thin-layer chromatography (TLC) counterpart. However, TLC allows for faster development times, yields better separation, and offers a wide range of adsorptive supports (*41*). Lea *et al.* (*78*) (using both cellulose- and silica-based supports) and Oszmianski and Sapis (*79*) have successfully used TLC to identify the dimeric procyanidins from seeds and wines.

Liquid-liquid partition chromatography has been successfully used to effect preliminary separation of procyanidins based upon rough molecular weight classes. Rossi and Singleton (60) separated grape seed extract into three fractions: an ether fraction containing the monomeric flavan-3-ols, an ethyl acetate fraction containing oligomeric material and an ethanol fraction containing higher polymeric material. Arnold et al. (62) used a similar partition system to fractionate seed tannins. The solvents ether, ethyl acetate, butanol and ethanol were used to elute the seed material from a diatomaceous earth column. The fractions obtained separated the procyanidins into the monomers (ether), the dimers (ethyl acetate), the trimers and tetramers (butanol), and the pentameric and larger polymers (ethanol) (80).

In 1972 Thompson et al. (81) published their seminal paper on separation and isolation of condensed tannins using low-pressure column chromatography on Sephadex LH-20 resin. Sephadex LH-20 is a counterpart to the G-25 dextran gel but with isopropyl groups linked into the dextran backbone, increasing hydrophobicity. The putative mechanism appears to be both through lone-pair interactions on the ether linkage with the pi system of the aromatic rings in the flavonoids as well as by hydrogen-bond capture within the resin itself until a suitable competitor is used to displace the material (e.g.–50% acetone). The Sephadex gel is typically prepared in 50% aqueous methanol; this solvent is also used to wash off carbohydrates and low molecular weight phenolics as it has been shown to be a poor eluant for the polymeric procyanidins (46).

Chromatography on Sephadex LH-20 is now the most widely used preparative technique, although other resins, such as Fractogel TSK ("Toyopearl") have been used either singly or in conjunction with Sephadex LH-20 (50, 82). Lea et al. (78) used Sephadex LH-20 along with counter-current distribution to separate dimeric and trimeric procyanidins. Oszmianski and Sapis (79) used Sephadex LH-20 as a semipreparative fractionation step in their study of grape seed phenolics. Kantz and Singleton (54, 58) used Sephadex LH-20 columns for separating the polymeric flavan-3-ol fraction from grape and wine extracts. The efficacy of the procedure was verified using the Folin Ciocalteau total phenolic colorimetric assay (83, 84).

As repetitive fractionations are often required to obtain pure compounds alternative approaches utilizing high performance liquid chromatography (HPLC) have been developed. HPLC has the benefits of increased rapidity, little to no sample pretreatment requirements, and (frequently) single run analysis (41, 42). Lea (85) found that the polymeric procyanidins material could be eluted from a reversed-phase column using a steep ramp in the organic phase during a gradient run. Despite the number of studies of grape and wine procyanidins using HPLC separation no group has effected separation of the procyanidins beyond the trimer fraction (86-88). Further separation of the "envelope" of polymeric procyanidins by HPLC appears to be quite difficult. Other complications involving HPLC include excessive tailing of peaks when nonacidified solvents are used, interflavan bond cleavage and rearrangement when acidified solvents are used, and metal complexation of the procyanidins from columns and frits (72, 73). Furthermore, as Singleton (41) has noted, HPLC chromatograms contain many peaks, the majority of which have yet to be identified, and preparative collection of these peaks for subsequent structural identification is a time-consuming process. With the advent of reliable HPLC-MS interfaces, however, this may no longer be a stumbling block.

Two more recent separation techniques which may prove invaluable in separating the procyanidins are centrifugal counter-current chromatography (CCC) and capillary electrophoresis (CE). Okuda's group in Japan has successfully used CCC in separating hydrolyzable tannins (76); Putman and Butler (89) have applied the technique to the fractionation of condensed tannins in sorghum grain. However, as Karchesy *et al.* (72) note, optimization of the solvent mixtures for high selectivity has proven difficult, and the technique at the moment appears relegated to a preliminary purification role. CE in neutral or charged mode (the phenolic hydroxyls are weak acids with pKa's of *ca.* 9) may prove the best approach to separation; however, research has yet to be done in this area. Regardless, it should be noted that analyzing for the polymers at elevated pH (>3) would necessitate precautions to prevent oxidation, which is greatly accelerated at more basic pH.

Subsequent structural determinations of the procyanidins isolated may be achieved by chemical degradation (using such compounds as toluene-alpha-thiol), ^1H and ^{13}C nuclear magnetic resonance (NMR), and fast-atom bombardment mass spectrometry (FAB-MS). Certainly n. O. e. NMR has proven to be of vast importance in structure elucidation (75). More recent work has focused on the fluorescent properties of the procyanidins in molecular weight determinations (90). These topics, however, are beyond the scope of this paper.

Conclusions

The flavan-3-ols and their polymers play a major role in wine taste. They are the most important contributors to bitterness and astringency. Whether or not the molecular weight of the polymeric form plays a role in bitter transduction has yet to be unambiguously ascertained; the increase in the number of hydroxyl groups as molecular weight increases , however, does appear to be the mechanism underlying astringency.

The most successful separations of the procyanidins to date have relied on low pressure column chromatography, typically on Sephadex LH-20, either alone or in combination with other resins, or in combination with HPLC. While the ease and rapidity of the latter technique is appealing, the problems associated with HPLC's use (metal complexation, interflavan bond cleavage) have yet to be resolved. With the advent of newer separation technologies (CCC, CE) the separation of the higher molecular weight oligomers may soon be accomplished.

Acknowledgments

The author would like to thank Dr. A. C. Noble and Dr. V. L. Singleton for graciously consenting to review this chapter, and for their helpful suggestions. This work was supported in part by funds provided by the American Society for Enology and Viticulture and by the Wine Spectator Scholarship Fund.

Literature Cited

1. Bartoshuk, L. M. *Food Tech.* **1991**, *45*, 108, 110, 111-112.
2. Nachman, M.; Cole, L. P. In *Chemical Senses, Part 2, Taste*; Beidler, L. M., Ed.; Handbook of Sensory Physiology IV; Springer-Verlag: Berlin, 1971; pp 337-362.
3. Avenet, P.; Lindemann, B. *J. Membrane Biol.* **1989**, *112*, 1-8.
4. Ramirez, I. *Neuro. Biobehav. Rev.* **1990**, *14*, 125-134.
5. Henkin, R. I.; Christiansen, R. L. *J. Applied Physiol.* **1967**, *22*, 316-320.
6. Plattig, K.-H. *Clin. Physics Physiol. Meas.* **1989**, *10*, 91-126.
7. Kinnamon, J. C. In *Neurobiology of Taste and Smell*; Finger, T. E.; Silver, W. L., Eds.; John Wiley & Sons: New York, 1987; pp 277-297.
8. Norgren, R. In *The Nervous System—Sensory Processes, Part 2*; Darian-Smith, I., Ed.; Handbook of Physiology; American Physiological Society: Bethesda, Maryland, 1984, Vol. 3; pp 1087-1128.
9. Beidler, L. M. *Prog. Biophys. Biophysical Chem.* **1961**, *12*, 107-151.
10. Collings, V. B. *Percept. Psychophys.* **1974**, *16*, 169-174.
11. Arvidson, K.; Friberg, U. *Science* **1980**, *209*, 807-808.
12. Kurihara, K.; Kashiwayanagi, M.; Yoshii, K.; Kurihara, Y. *Comm. Agric. Food Chem.* **1989**, *2*, 1-50.
13. Faurion, A. In *Progress in Sensory Physiology*; Ottoson, D., Ed.; Springer-Verlag: Berlin, 1987, Vol. 8; pp 129-201.
14. Schiffman, S. S.; Erickson, R. P. *Neuro. Biobehav. Rev.* **1980**, *4*, 109-117.
15. Schiffman, S. S.; Cahn, H.; Lindley, M. G. *Pharm. Biochem. Behav.* **1981**, *15*, 377-388.
16. McBurney, D. H. *Chem. Senses* **1974**, *1*, 17-27.
17. Pfaffmann, C.; Frank, M.; Norgren, R. *Ann. Rev. Psych.* **1979**, *30*, 283-325.
18. Faurion, A.; Vayssettes-Courchay, C. *Brain Res.* **1990**, *512*, 317-332.
19. Erickson, R. P. In *Taste, Olfaction and the Central Nervous System*; Pfaff, D. W, Ed.; Rockefeller University Press: New York, 1984; pp 129-150.
20. Frank, M. E. In *Taste, Olfaction and the Central Nervous System*; Pfaff, D. W, Ed.; Rockefeller University Press: New York, 1984; pp 107-128.
21. Kinnamon, S. C. *Trends Neuro.* **1988**, *11*, 491-496.
22. Shallenberger, R. S.; Acree, T. E. In *Chemical Senses, Part 2, Taste*; Beidler, L. M., Ed.; Handbook of Sensory Physiology IV; Springer-Verlag: Berlin, 1971; pp 221-277.
23. Shepherd, G. M. *Cell* **1991**, *67*, 845-851.
24. Kurihara, K.; Kamo, N.; Kobatake, Y. In *Advances in Biophysics*; Kotani, M., Ed.; University Park Press: Baltimore, MD, 1978, Vol. 10; pp 27-95.
25. Lawless, H. T. In *Neurobiology of Taste and Smell*; Finger, T. E.; Silver, W. L., Eds.; John Wiley & Sons: New York, 1987; pp 401-420.
26. Haslam, E.; Lilley, T. H. *Crit. Rev. Food Sci. Nutr.* **1988**, *27*, 1-40.
27. Joslyn, M. A.; Goldstein, J. L. In *Advances in Food Research*; Chichester, C. O.; Mrak, E. M.; Stewart, G. F., Eds.; Academic Press: New York, 1964, Vol. 13; pp 179-217.

28. Amerine, M. A.; Ough, C. S. *Methods for Analysis of Musts and Wines*; John Wiley & Sons: New York, 1980.
29. Noble, A. C.; Bursick, G. F. *Am. J. Enol. Vit.* **1984**, *35*, 110-112.
30. Amerine, M. A.; Roessler, E. B. *Wines, Their Sensory Evaluation*; W. H. Freeman: New York, 1983.
31. Noble, A. C.; Philbrick, K. C.; Boulton, R. B. *J. Sens. Studies* **1986**, *1*, 1-8.
32. Singleton, V. L.; Noble, A. C. In *Phenolic, Sulfur, and Nitrogen Compounds in Food Flavors*; Charalambous, G.; Katz, I., Eds.; ACS Symposium Series; ACS: Washington, D.C., 1976, Vol. 26; pp 47-70.
33. Singleton, V. L.; Esau, P. *Phenolic Substances in Grapes and Wines, and their Significance*; Advances in Food Research; Academic Press: New York, 1969; Supplement 1.
34. Van der Wel, H.; Van der Heijden, A.; Peer, H. G. *Food Rev. Int.* **1987**, *3*, 193-268.
35. Nishimura, T.; Kato, H. *Food Rev. Int.* **1988**, *4*, 175-194.
36. Vérette, E.; Noble, A. C.; Somers, T. C. *J. Sci. Food Agric.* **1988**, *45*, 267-272.
37. Crespo-Riera, G. E. *Specific Bitterness in Muscat and Gewurztraminer Dry Table Wines*; M.S. Dissertation; University of California: Davis, CA, 1986.
38. Noble, A. C.; Strauss, C. R.; Williams, P. J.; Wilson, B. *Am. J. Enol. Vit.* **1988**, *39*, 129-131.
39. Su, C. T.; Singleton, V. L. *Phytochem.* **1969**, *8*, 1553-1558.
40. Singleton, V. L. In *Grape and Wine Centennial Symposium*; Webb, A. D., Ed.; University of California: Davis, CA, 1980; pp 215-227.
41. Singleton, V. L. In *Wine Analysis*; Linskens, H. F.; Jackson, J. F., Eds.; Modern Methods of Plant Analysis; Springer-Verlag: Berlin, 1988, Vol. 6; pp 173-218.
42. Somers, T. C.; Vérette, E. In *Wine Analysis*; Linskens, H. F.; Jackson, J. F., Eds.; Modern Methods of Plant Analysis; Springer-Verlag: Berlin, 1988, Vol. 6; pp 219-257.
43. Romeyer, F. M.; Macheix, J.-J.; Sapis, J.-C. *Phytochem.* **1986**, *25*, 219-221.
44. Ribéreau-Gayon, P. *Plant Phenolics*; University Reviews in Botany; Hafner Publishing Company: New York, 1972, Vol. 3.
45. Czochanska, Z.; Foo, L. Y.; Newman, R. H.; Porter, L. J.; Thomas, W. A. *J. Chem. Soc. Chem. Comm.* **1979**, 375-377.
46. Foo, L. Y.; Porter, L. J. *Phytochem.* **1980**, *19*, 1747-1754.
47. Jacques, D.; Haslam, E.; Bedford, G. R.; Greatbanks, D. *J. Chem.Soc. Perkin Trans. 1* **1974**, 2663-2671.
48. Karchesy, J. J.; Hemingway, R. W. *J. Agric. Food Chem.* **1986**, *34*, 966-970.
49. Hemingway, R. W. In *Chemistry and Significance of Condensed Tannins*; Hemingway, R. W.; Karchesy, J. J., Eds.; Plenum Press: New York, 1989; pp 83-107.
50. Ricardo da Silva, J. M.; Rigaud, J.; Cheynier, V.; Cheminat, A.; Moutounet, M. *Phytochem.* **1991**, *30*, 1259-1264.

51. Hemingway, R. W.; Foo, L. Y.; Porter, L. J. *J. Chem. Soc. Perkin Trans. 1* **1982,** 1209-1216.
52. Glories, Y. In *Plant Flavonoids in Biology and Medicine*; Cody, V.; Middleton, E. J.; Harborne, J. B.; Beretz, A., Eds.; Alan R. Liss: Strasbourg, France, 1987; pp 123-134.
53. Cork, S. J.; Krockenberger, A. K. *J. Chem. Ecol.* **1991,** *17,* 123-134.
54. Kantz, K.; Singleton, V. L. *Am. J. Enol. Vit.* **1990,** *41,* 223-228.
55. Meyer, J.; Hernandez, R. *Am. J. Enol. Vit.* **1970,** *21,* 184-188.
56. Singleton, V. L.; Draper, D. E. *Am. J. Enol. Vit.* **1964,** *15,* 34-40.
57. Singleton, V. L. In *Plant Polyphenols; Synthesis, Properties, Significance*; Hemingway, R. W.; Laks, P. E., Eds.; Plenum Press: New York, 1992; pp 859-880.
58. Kantz, K.; Singleton, V. L. *Am. J. Enol. Vit.* **1991,** *42,* 309-316.
59. Singleton, V. L.; Trousdale, E. K. *Am. J. Enol. Vit.* **1992,** *43,* 63-70.
60. Rossi, J. A., Jr.; Singleton, V. L. *Am. J. Enol. Vit.* **1966,** *17,* 240-246.
61. Delcour, J. A.; Vandenberghe, M. M.; Corten, P. F.; Dondeyne, P. *Am. J. Enol. Vit.* **1984,** *35,* 134-136.
62. Arnold, R. A.; Noble, A. C.; Singleton, V. L. *J. Agric. Food Chem.* **1980,** *28,* 675-678.
63. Leach, E. J. *Evaluation of Astringency and Bitterness by Scalar and Time-Intensity Procedures*; M.S. Dissertation; University of California: Davis, CA, 1985.
64. Robichaud, J. L.; Noble, A. C. *J. Sci. Food Agric.* **1990,** *53,* 343-353.
65. Lea, A. G. H.; Arnold, G. M. *J. Sci. Food Agric.* **1978,** *29,* 478-483.
66. Arnold, R. A.; Noble, A. C. *Am. J. Enol. Vit.* **1978,** *29,* 150-152.
67. Lea, A. G. H. In *Bitterness in Food and Beverages*; Rousseff, R., Ed.; Elsevier: New York, 1990; pp 123-143.
68. Larson-Powers, N.; Pangborn, R. M. *J. Food Sci.* **1978,** *43,* 41-46.
69. Fischer, U. *The Influence of Ethanol, pH, and Phenolic Composition on the Temporal Perception of Bitterness and Astringency, and Parotid Salivation*; M.S. Dissertation; University of California: Davis, CA, 1990.
70. Singleton, V. L.; Sieberhagen, H. A.; de Wet, P.; van Wyk, C. J. *Am. J. Enol. Vit.* **1975,** *26,* 62-69.
71. Guinard, J.-X.; Pangborn, R. M.; Lewis, M. J. *Am. J. Enol. Vit.* **1986,** *37,* 184-189.
72. Karchesy, J. J.; Bae, Y.; Chalker-Scott, L.; Helm, R. F.; Foo, L. Y. In *Chemistry and Significance of Condensed Tannins*; Hemingway, R. W.; Karchesy, J. J., Eds.; Plenum Press: New York, 1989; pp 139-151.
73. Karchesy, J. J. In *Chemistry and Significance of Condensed Tannins*; Hemingway, R. W.; Karchesy, J. J., Eds.; Plenum Press: New York, 1989; pp 197-202.
74. Barofsky, D. F. In *Chemistry and Significance of Condensed Tannins*; Hemingway, R. W.; Karchesy, J. J., Eds.; Plenum Press: New York, 1989; pp 175-195.

75. Ferreira, D.; Brandt, E. V. In *Chemistry and Significance of Condensed Tannins*; Hemingway, R. W.; Karchesy, J. J., Eds.; Plenum Press: New York, 1989; pp 153-173.
76. Okuda, T.; Yoshida, T.; Hatano, T. *J. Nat. Products* **1989**, *52*, 1-31.
77. Singleton, V. L.; Draper, D. E.; Rossi, J. A., Jr. *Am. J. Enol. Vit.* **1966**, *17*, 206-217.
78. Lea, A. G. H.; Bridle, P.; Timberlake, C. F.; Singleton, V. L. *Am. J. Enol. Vit.* **1979**, *30*, 289-300.
79. Oszmianski, J.; Sapis, J. C. *J. Agric. Food Chem.* **1989**, *37*, 1293-1297.
80. Watson, B. T. *Anthocyanin Tannin Interactions During Table Wine Aging*; M.S. Dissertation; University of California: Davis, CA, 1975.
81. Thompson, R. S.; Jacques, D.; Haslam, E.; Tanner, R. J. N. *J. Chem. Soc. Perkin Trans. 1* **1972**, 1387-1399.
82. Ricardo da Silva, J. M.; Rosec, J.-P.; Bourzeix, M.; Heredia, N. *J. Sci. Food Agric.* **1990**, *53*, 85-92.
83. Slinkard, K.; Singleton, V. L. *Am. J. Enol. Vit.* **1977**, *28*, 49-55.
84. Singleton, V. L.; Rossi, J. A., Jr. *Am. J. Enol. Vit.* **1965**, *16*, 144-158.
85. Lea, A. G. H. *J. Chromat.* **1980**, *194*, 62-68.
86. Nagel, C. W. *Cereal Chem.* **1985**, *62*, 144-147.
87. Roggero, J.-P.; Coen, S.; Archier, P. *J. Liq. Chromat.* **1990**, *13*, 2593-2603.
88. Revilla, E.; Bourzeix, M.; Alonso, E. *Chromatographia* **1991**, *31*, 465-468.
89. Putman, L. J.; Butler, L. G. *J. Chromat.* **1985**, *318*, 85-93.
90. Cho, D.; Mattice, W. L.; Porter, L. J.; Hemingway, R. W. *Polymer* **1989**, *30*, 1955-1958.

RECEIVED February 9, 1993

Chapter 5

Progress in Beer Oxidation Control

Nick J. Huige

Miller Brewing Company, 3939 West Highland Boulevard,
Milwaukee, WI 53201

Beer flavor instability is caused by the formation of volatile, long chain, unsaturated carbonyls with low flavor thresholds and unpleasant flavors. Long chain unsaturated aldehydes, such as trans-2-nonenal which contributes a cardboard-like flavor, are prime contributors.

Five main mechanisms for the formation of volatile, long chain, unsaturated carbonyls are discussed: 1) Strecker degradation of amino acids; 2) melanoidin mediated oxidation of higher alcohols; 3) oxidative degradation of iso-alpha-acids; 4) aldol condensation of short chain aldehydes, and, 5) enzymatic or non-enzymatic oxidation of fatty acids. Various mechanisms are most likely involved including the enzymatic degradation of fatty acids during malting and mashing followed by auto-oxidation of intermediates in the brewhouse to precursors which are oxidized in the package under the influence of free radical forms of oxygen.

Various methods to minimize oxidation reaction are discussed in detail. They are: 1) maintaining a high reducing potential through oxidation control in the brewhouse; 2) minimizing oxygen pickup by the product in cellar operation, during packaging, and during product storage; 3) minimizing oxygen radicals through optimal use of endogenous and exogenous antioxidants and through minimizing copper and iron pickup; and, 4) avoidance of high storage temperatures.

Beer flavor is never constant. During the aging process subtle changes occur that round out the beer flavor and reduce some of the sulphury notes associated with freshly fermented beer. These beneficial changes often continue for several weeks

0097–6156/93/0536–0064$09.50/0

after packaging *(1)* at which point most American lager beers reach their optimum flavor. The changes in flavor that occur thereafter are undesirable for American lager beers, but may actually be preferred for some of the heavier European specialty beers. An excellent overview of flavor changes during storage of beer, wine, and cider is given by Laws and Peppard *(2)*.

Flavor changes in packaged lager beers go through a number of stages *(3)*. The rate of change is strongly dependent on storage temperature. During the initial stage (after one to three months at room temperature) a papery, or cardboard-like flavor develops and the product decreased in fruity/estery and floral character. During the next phase the beer becomes bready, sweeter and toffee-like and some beers develop straw-like, earthy, and sometimes metallic flavors. In the final stage woody or leathery flavors develop and the beer attains a winey, sherry-like character. During this entire process most products decrease in bitterness *(4)*.

Different beers age in different ways *(5, 6)*. For example, ales become distinctly sweet with a molasses-like, cloying character. Stouts often develop stale, cheesy characteristics. In general, darker, heavier beers have a better flavor stability. Besides, because of their greater flavor strength, changes in certain components are less noticeable.

Chemically, the flavor changes that develop are very hard to define since they change progressively during storage and are due to a combination of the effects of many flavor components rather than only one or a few *(1)*. A better understanding of flavor instability is further hampered by the fact that flavor changes are not only dependent on the type of beer, but also on brewing procedures, raw materials, packaging conditions and environmental factors. It is no surprise therefore that flavor instability is the most widely researched topic in brewing science. Many excellent reviews have been published *(1, 2, 4, 7, 8)* regarding various aspects of flavor instability.

It is generally believed that oxidation is the principal cause of flavor instability. Oxidation can take place during malting, in the brewhouse, or during aging in the brewery or in the package. The problem is that the effect of these oxidation reactions is often not seen until many weeks after packaging. Most of the oxidative flavor changes that take place are the result of the formation of volatile carbonyls.

The review below will first focus on the most important volatile carbonyls and possible mechanisms for their formation. Next, changes in wort and beer processing and packaging will be discussed that have been proposed to reduce beer oxidation. A number of graphs will also be presented to illustrate the relative importance of package oxygen and beer storage temperature.

Importance of Volatile Carbonyls

Carbonyl compounds formed or released during the storage of packaged beer are said to be the main cause of oxidized flavor development (1, 7, 9). Hashimoto (7) showed an excellent correlation between the sensory score for oxidized flavor and the concentration of volatile aldehydes. It has been shown (10) that after a carbonyl scavenger, such as hydroxylamine or 2,4-dinitrophenylhydrazine was added to oxidized beer, the oxidized flavor disappeared within minutes.

Volatile aldehydes in beer generally have unpleasant flavors and they become more unpleasant as their carbon number increases. Table 1 gives some saturated and unsaturated aldehydes that have been detected in beer along with their typical flavor and threshold level in beer (11-14). It can be seen that threshold levels generally decrease with increasing chain lengths. Carbonyls that are present in beer at concentrations below their threshold may nevertheless be important because of synergistic flavor contributions. Meilgaard (11) showed that the threshold level of a carbonyl mixture was exceeded when four carbonyls were mixed in beer which were each present below their individual threshold levels.

Table 1 also indicates the concentrations of these aldehydes in old beer (13, 14) and whether or not their concentrations increased during storage in the package. Trans-2-nonenal which has a papery, cardboard-like flavor and aroma has a threshold level of only 0.1 ppb. Trans-2-nonenal is an important carbonyl compound since it is most often associated (15-20) with the oxidative flavor which develops in lager beer during the first three months of storage. Of all the aldehydes in Table 1, trans-2-nonenal is the only aldehyde that increases during storage to above threshold levels.

Dalgliesh (4) has commented that many investigations on trans-2-nonenal are done with heated beer (13, 14) to accelerate aging or with acidified heated beer (17, 21) to maximize nonenal production. He wonders whether under normal storage conditions trans-2-nonenal would play as dominant a role as many investigators think. According to Greenhoff (13), however, the type of cardboard flavor that results after accelerated aging at 60°C is similar to those that develop at 37°C or at 18°C over longer periods of time. Generally all investigators agree that trans-2-nonenal is an important component in the complex mixture of oxidized beer flavors, although certainly many other components contribute as well.

Besides the carbonyls shown in Table 1, there are many other carbonyls in beer including additional aldehydes and many ketones. Many of these carbonyls are present at sub-threshold levels even though their concentrations are often many orders of magnitude higher than the aldehydes shown in Table 1. Some of these

Table 1. Carbonyls in Beer

Aldehydes	Typical Flavor Notes	Increase with Age	Threshold Value (ppb)	Conc. in Aged Beer (ppb)
Alkanels				
Pentanal	Grassy	Yes	500	6
Hexanal	Vinous, Bitter	Yes	300	4
Heptanal	Vinous, Bitter	Yes	50	2
Octanal	Vinous, Bitter	Yes	40	1.7
Nonanal	Astringent	No	15	4
Decanal	Bitter	No	5	1
Alkenals				
2-Hexenal	Astringent	Yes	500	1
2-Octenal	Bitter, Stale	Yes	0.3	0.14
2-Nonenal	Paper, Cardboard	Yes	0.1	0.16-0.48
2-Decenal	Bitter, Rancid	Yes	1.0	0.1
Alkadienals				
2,4-nonadienal	Rancid	No	0.05	--
2,6-nonadienal	Cucumber	No	0.5	0.7
2,4-decadienal	Oily, Rancid	No	0.3	0.8

carbonyls compete with each other in the oxidative reaction scheme during the storage of beer, and sometimes their concentrations increase in parallel with the oxidative flavor development even though they are not responsible for it. Compounds such as acetaldehyde *(9,22)*, furfural *(23, 24)* and hydroxy methylfurfural *(4)*, and **gamma-hexalacton, gamma-nonalacton, 3-methyl butanal,** heptanal, and nicotinic acid ethyl ester *(23)* appear to be useful indicators for beer oxidation. Some of these indicators are very temperature dependent while others are sensitive to package oxygen levels *(3)*.

Mechanisms of Beer Oxidation

Since volatile carbonyls appear to be the main contributors to the oxidative flavor that develops in packaged beer during storage, it will be no surprise that much research effort has been devoted to finding out how these carbonyls are formed, where they are formed, and how their formation can be minimized by technological improvements.

Five main mechanisms of volatile carbonyl formation have been proposed:

1. Strecker Degradation of Amino Acids
2. Melanoidin Medicated Oxidation of Higher Alcohols
3. Oxidative Degradation of Iso-Alpha-Acids.
4. Aldol Condensation of Short Chain Aldehydes into Longer Ones
5. Enzymatic or Non-Enzymatic Degradation of Fatty Acids

Each one of these mechanisms will be discussed below along with critiques of some investigators. Very likely, several of these mechanisms will be taking place at the same time, the dominant mechanisms being dependent on product, product history, and storage conditions.

It will be shown that there is no agreement among investigators where the major volatile aldehydes such as trans-2-nonenal are formed and how they end up in the final product during storage. Major difficulties in staling research have been encountered in developing sensitive techniques to detect volatile aldehydes in sub-ppb quantities and to follow their fate and that of their precursors throughout the process and during storage. Volatile unsaturated aldehydes, for example, can exist in equilibria with their less obnoxious hydrated forms and as flavor inactive complexes with SO_2. These equilibria can shift as a result of temperature or pH changes or due to changes in the concentrations of other aldehydes and carbonyls that are often present in concentrations that are orders of magnitude higher than the unsaturated volatile aldehydes of interest. Investigators have therefore resorted to the use of model systems *(7)* or to the use of more encompassing predicting measures such as nonenal potential *(16)*.

1.   **Strecker Degradation of Amino Acids** Several investigators *(9, 25, 26)*
        have proposed that amino acids in beer can be degraded by the Strecker
        reaction to give aldehydes with the same carbon skeleton as the amino acid
        but without one carbon at the place of the carboxylic group. Blockmans
        (25) demonstrated the formation of isobuteraldehyde from valine and
        isovaleraldehyde from leucine in beer. The formation of formaldehyde
        from glycine and acetaldehyde from alanine were also demonstrated in a
        model beer solution *(26)*.

        It can be concluded than even though formation of these saturated
        aldehydes by this mechanism is possible, oxidative flavor formation as a
        result of these aldehydes is unlikely as their flavor thresholds are quite high.
        Also, trans-2-nonenal cannot be formed by Strecker degradation, since a
        natural amino with appropriate side chain does not exist *(27)*.

2.   **Melanoidin Mediated Oxidation of Higher Alcohols** Hashimoto and
        co-workers *(7, 26, 28, 29)* are avid proponents of a mechanism in which
        alcohols in beer can undergo oxidation to give the corresponding
        aldehydes, the oxidant being the carbonyl groups present in melanoidins.
        Melanoidins are formed by the amino-carbonyl reaction during the boiling
        of wort and kilning of malt. In this mechanism, therefore, oxidized
        melanoidins accept hydrogen atoms from alcohols with the formation of
        corresponding aldehydes. Molecular oxygen does not oxidize alcohols
        directly in the absence of melanoidins, but the presence of oxygen
        accelerates the transformation of the reactive carbonyl groups of
        melanoidins involved in the electron transfer system. The formation of
        volatile aldehydes by this mechanism was shown *(26)* in a model beer
        solution that was stored at 50°C. The oxidation of alcohols increased with
        increasing melanoidin concentration and proceeded more rapidly at higher
        temperatures, higher air levels and lower pH values.

        The relevance of the alcohol oxidation mechanism was further
        demonstrated by Hashimoto *(29)* by storage experiments with hopped wort
        and beer. It was found that wort, which does not contain higher alcohols,
        did not develop the typical stale flavor during storage, whereas beer did.
        This difference, however, may have been partly due to the higher pH of
        wort.

        Isohumulones and polyphenols can also oxidize higher alcohols to
        aldehydes through the action of their phenoxy radicals. However, in a
        reaction system such as beer where isohumulones, polyphenols and
        melanoidins are all present, isohumulones and polyphenols are more likely

to donate hydrogen atoms preferentially to melanoidins, rather than accepting hydrogen from the alcohols *(7, 26)*.

Devreux and co-workers *(1)* demonstrated that the oxidation of alcohols in beer is very slow in the darkness. In the presence of light and riboflavin, the reaction is fast, but it is greatly inhibited by low concentrations of polyphenols. On the other hand, as alcohols are less reactive when their molecular weight is high and as ethanol is the most abundant alcohol in beer, it seems unlikely that the formation of long chain volatile aldehydes from higher alcohols is a very important mechanism. Barker and co-workers *(27)* came to a similar conclusion. In a model system containing 2-nonen-1-ol and 5% ethanol Irwin and co-workers *(41)* found that only 0.2% of the 2-nonen-1-ol was oxidized to 2-nonenal. Based on this low yield and the low concentration of 2-nonen-1-ol in beer of less than 0.5 ppb, they concluded that it is unlikely that 2-nonenal is produced by this mechanism.

3.   **Oxidative Degradation of Iso-Alpha-Acids** Hashimoto and Eshima *(30)* found that unhopped beers hardly ever develop a typical oxidized flavor on storage and suggested that isohumulones play an important role in the flavor staling of beer. They designed an experiment in which higher alcohols obtained from beer and isohumulones were added to a model melanoidin solution at pH 4.2 which was subsequently stored at 40°C for 10 days. About 0.01% of the higher alcohols and 3% of the isohumulones were oxidized to volatile carbonyls. The oxidation products from isohumulones were alkenals and alkadienals with chain lengths of C6 through C12. These products formed more rapidly when higher alcohols were also present.

Hashimoto *(29, 32)* claims that because the double bonds or carbonyl group of the isohexenoyl side chain of isohumulones are involved in the oxidative degradation of isohumulones, the use of tetrahydroisohumulones or rho-isohumulones as bittering agents instead of isohumulones effectively prevented the development of stale flavor, at least in their beers.

Barker and co-workers *(27)* commented that the degradation of isohumulones cannot form the important trans-2-nonenal since appropriate size side chains do not exist. On the other hand, trans-2-nonenal may be formed by aldol condensation (see below) of heptanal and acetaldehyde.

One possible way that iso-alpha-acids can oxidize is by donating electrons to melanoidins *(31)*. This in fact inhibits the melanoidin mediated oxidation of alcohols discussed above. Thus, by reducing the melanoidins, the iso-alpha-acids are in fact protecting beer flavor.

The oxidative reactions described above pertain to fresh hops. During the storage of hops, alpha-acids can also oxidize to far more unpleasant compounds (31). For example, by cleavage of the acyl side chain in alpha-acids, fatty acids such as isovaleric, isobuteric, and 2-methyl buteric acids may be formed, which have very stale, cheesy flavors.

4. **Aldol Condensation of Short Chain Aldehydes into Longer Ones**
Aldol condensation of aldehydes takes place under the mild conditions existing in beer during shelf storage (7). Aldol condensation was studied by Hashimoto and Kuroiwa (26) in a small model beer system containing 20 mmol of proline and storage conditions of 20 days at a rather high temperature of 50°C. They found that 2-butanal was formed from acetaldehyde, while 2-pentyl-2-butanal and trans-2-nonenal were formed from acetaldehyde and heptanal. Acetaldehyde and n-butanal formed various other condensation products. Proline or other amino acids appear to be essential catalysts in the aldol condensation reaction as it is utilized in the formation of imine intermediates.

From these findings, it was concluded that it is likely that the higher alkenals and alkedienals that contribute to the stale flavor of beer may be derived through aldol condensation of shorter chain aldehydes formed by the Strecker degradation of amino acids, oxidation of higher alcohols, or oxidative degradation of isohumulones. It should be noted, however, that in the experiments with model beer solutions only 1-2 ppm of 2-nonenal was found after the addition of 500 ppm acetaldehyde and 500 ppm n-heptanal. The approximate yield is therefore only 0.3%. Since the concentration of n-heptanal in beer is approximately 1.2 ppb, a reaction yield of 0.3% would form 2-nonenal at concentrations considerably below threshold.

5. **Enzymatic or Non-Enzymatic Oxidation of Fatty Acids** The majority of investigators agree that lipids from malt are a major source of volatile stale aldehydes. Lipids in malt are broken down by lipase to fatty acids which provide much of the energy required for embryo growth during the early stages of malting. Following is a brief review of the proposed mechanisms of fatty acid oxidation and a description of how volatile aldehydes or intermediates may increase or decrease as they proceed through the brewing process.

5a.   **Enzymatic Oxidation of Unsaturated Fatty Acids** Unmalted
barley contains a lipoxygenase enzyme system that is capable of
oxidizing linoleic acid into 9-hydroperoxy-trans-10, cis-12-
octadecadienoic acids (9-L00H). During germination, an iso
enzyme is formed which transforms linoleic acid into a
corresponding 13-hydroperoxy acid *(33)*. These hydroperoxy-acids
can be degraded under the influence of heat or metal catalysts into
flavor active volatile aldehydes and oxy-acids, or they may be
reduced to the corresponding 9- and 13-hydroxy acids *(34)*.

Hydroperoxy acids can also be transformed into α or gamma-ketols
under the influence of isomerase which is formed during
germination. Through a series of chemical reactions catalyzed by
heat and light, these ketols are transferred into mono-, di-, or trihydroxy acids which can be found in beer in ppm quantities. By
thermal degradation during kilning or wort boiling hydroxy acids,
especially the vicinal di-hydroxy acids are converted into hexanal
and 2-nonenal.

Drost and co-workers *(21)* also concluded that 9, 10, 13-
trihydroxy, 11-trans-octadecanoic acid was the most important
precursor of the cardboard stale flavor. Addition of 3 ppb to beer
and heating to 60°C for 24 hours caused a significant increase of
the cardboard flavor. Reducing the pH to 2 accelerated this flavor
formation. Other trihydroxy acids and their isomers were also
implicated as likely precursors.

The results obtained by heating beer to 60°C and lowering the pH
could lead to false conclusions, since van Eerde and Strating *(35)*
failed to find a correlation between trans-2-nonenal formed by
heating beer to 40°C and trans-2-nonenal formed by prolonged
storage of beer at room temperature.

Stenroos and co-workers *(36)* showed that trihydroxy acids were
not able to produce trans-2-nonenal at a beer pH of 4, even at
temperatures of 38°C.

Domingues and Canales *(37)* concluded that trans-2-nonenal is
formed from linoleic or linolenic acid by enzymatic and/or free
radical oxidative degradations, possibly involving colneleic and
colnelenic acids as intermediates. Garza-Ulloa and co-workers *(38)*
showed that the addition of colneleic acid to package beer and
storage at 32°C for three days increased trans-2-nonenal and

produced a cardboard flavor, which did not occur when colneleic acid was not added. However, colneleic acid has not been identified to exist in beer.

There is no doubt that enzymatic degradation of fatty acids can produce trans-2-nonenal and many other volatile unsaturated aldehydes. Tressl and co-workers *(34)* measured 32 alkehydes in pale and dark colored malts, many of them in concentrations that would exceed their threshold levels in beer by orders of magnitude. Trans-2-nonenal, for example, was present in concentrations of 200 to 400 ppb. During mashing and boiling in the brewhouse, a substantial amount of carbonyl stripping takes place, but volatile carbonyls may also be formed at the same time through enzymatic (mash tun) and/or thermal (mash tun or brew kettle) degradation of fatty acids or intermediates. Trans-2-nonenal levels in wort going to fermentation are about 1 to 2 ppb.

Lipoxygenase is present in barley and increases in amount during malting *(39)*. During kilning lipoxygenase is largely destroyed and only a small amount of activity remains during the initial stages of mashing *(40)*. No activity is left after heating to 70°C at the end of mashing. Hydroperoxide isomerase is more heat stable than lipoxygenase, but it is not extracted in the wort under normal circumstances, according to Schwarz and Pyler *(40)*. Lipoxygenase and hydroperoxide isomerase activities may vary significantly due to barley variety and growth location.

Some of the more heat stable forms of lipoxygenase and hydroperoxide isomerase might actually result from the microflora that is present during malting, as these enzymes from microorganisms are generally more stable than malt derived enzymes.

Drost and co-workers *(16)* have shown that malt with a higher lipoxygenase activity has a higher nonenal potential. Products with a high nonenal potential developed more stale flavor.

5b.   **Auto-oxidation of Unsaturated Fatty Acids** Auto-oxidation is a free radical mechanism of oxidation in which a hydrogen atom is removed from the fatty acid molecule. The resulting radical reacts with molecular oxygen to form a peroxy radical, which in turn, removes a hydrogen atom from another fatty acid while forming a hydroperoxide *(7)*. This hydroperoxide is polymerized or it is cleaved to lower fatty acids or to aldehydes or other compounds.

The auto-oxidation mechanism was proposed and discussed by a number of investigators *(1, 7, 16, 19, 20, 37)*. Visser and Lindsay *(20)* suggested a free radical reaction mechanism involving hydroperoxide radicals in which trans-2-nonenal and other flavor active aldehydes are formed in packaged beer at normal temperature and pH from ethyl esters of free fatty acids formed by fermentation.

Free radical reactions may be initiated through the action of active forms of oxygen. For example, Kaneda and co-workers *(42)* showed that the degradation of fatty acids in beer at 60°C was accelerated by additions of Fe^{2+} and/or hydrogen peroxide. Clarkson and co-workers *(43)* showed that the activation of the oxygen molecule to peroxide might take place in the mash tun through the action of peroxidase. Activation of oxygen to superoxide and other radical forms might also happen through the action of enzymes such as polyphenol oxidase and lipoxygenase.

Irwin and co-workers *(41)* also conclude that active forms of oxygen may be involved in producing unsaturated aldehydes in beer from unsaturated hydroxy fatty acids through a metal catalyzed oxidation. Saturated hydroxy fatty acids or unsaturated fatty acids that are not yet oxidized are much less reactive at normal beer storage conditions and will not yield oxidation degradative products. However, in the brewhouse at much higher temperatures, metal catalyzed auto-oxidation of these compounds may occur.

Carbonyl Complex Equilibria

Volatile carbonyls may exist in malt, wort, or beer in equilibrium with their flavor inactive complexed forms. It has been shown that the addition of precursors of volatile carbonyls to finished beer does not always produce increased levels of volatile carbonyls in the product nor does it always produce the expected increase in stable flavor at normal temperatures and pH. It was postulated by several investigators *(16, 23-29, 41, 44, 45)* that many volatile carbonyls, including trans-2-nonenal are already present in the beer as it is packaged, but that they are present as flavor inactive complexed forms. During storage the volatile carbonyls may be slowly released from their complexes through shifts in equilibrium.

Gracey and co-works *(45)* have shown that 2-alkenals in beer and wort may co-exist with their hydrated forms as 3-hydroxy alkanals throughout the brewing process. This hydration was found to occur to considerable extent and the proportion of the hydrated compound at apparent equilibrium increased with the chain length. Their results with model systems indicate that the equilibrium for hydration would be substantially approached during the period of elevated temperature during wort production and certainly during the longer steps in fermentation and maturation.

Barker and co-workers *(27)* showed that bisulphite added to model beer solutions can form flavor inactive complexes with saturated or unsaturated aldehydes. They found that saturated aldehydes bind bisulphite more strongly than unsaturated aldehydes and that short chain aldehydes bind bisulphite more strongly than longer chain aldehydes. In beer, bisulphite addition complexes can exist with many of the aldehydes that are present. The complex formation is reversible and any decrease in sulphite concentration would increase the concentration of free aldehydes. Aldehydes and other carbonyls also compete for available sulfite and may displace each other from their complexes. For example, acetaldehyde formed by oxidation of ethanol during storage may displace trans-2-nonenal from its SO_2 complex as it binds more strongly to SO_2 than trans-2-nonenal does.

Bottled beers with a low SO_2 content developed a typical cardboard oxidized flavor more quickly than beers with a higher SO_2 level, according to Nordlöv and Winell *(44)*. Dufour *(79)*, however, is of the exact opposite opinion. He suggests that SO_2 formed during fermentation could carry a high level of carbonyls into the product as unstable bisulphite adducts, which could increase the susceptibility of that beer to early staling.. Nonenal and other aldehydes are released on equilibrium changes caused by: 1) oxidative loss of bisulphite; 2) binding of bisulphite by acetaldehyde that is formed; 3) increased temperature; or, 4) lower pH. This may explain why some of the forced tests, in which beer is heated to 60°C or its pH is artificially lowered, are not good predictors for stale flavor formation.

Both bound and free SO_2 has been found to decrease during storage of packaged beer *(27)*. Possible mechanisms for this decrease include: 1) the oxidation to sulphate; 2) addition to oxidized polyphenols; 3) addition to unsaturated sugar-amino acid reaction products; or, 4) reaction with protein disulphide bonds.

Bisulphite complexes have been found *(41)* to be susceptible to Cu(II) - catalyzed oxidation. Brown *(46)* found that the drop in bisulphite level is accompanied by a drop in the thiol level. He proposed a mechanism in which two reduced protein thiols are oxidized to form a protein disulphide. This disulphide in turn reacts with bisulphite to yield one molecule of thiosulphate and one thiol. The net effect is an oxidative conversion of a thiol and bisulphite to a thiosulphate. Since there are about 50 to 70 ppm of non-volatile organic sulphur compounds in beer, there is ample material available to remove bisulphites.

Even though the release of volatile unsaturated aldehydes from their SO_2 complexes is able to explain the general formation of stale flavor during storage, it does not necessarily mean that this is the main mechanism. Drost and co-workers *(16)* conclude that it is not very likely that saturated or unsaturated aldehydes that are present in wort will form bisulphite complexes during fermentation, since this reaction will compete with the reduction of these aldehydes to alcohols by yeast and since SO_2 formation during fermentation does not start until the alcohol level is about 1.5% . On the other hand, the small amount of SO_2 that is present in pitching yeast at the start of fermentation might be sufficient to bind the very low but significant levels of volatile long chain aldehydes that remain after wort cooling.

The reduction of aldehydes and ketones during fermentation was studied by a number of investigators *(16, 47)*. The reaction appears to be quite rapid. A wide range of aldehydes added to wort during or prior to fermentation, including trans-2-nonenal, were reduced quantitatively by yeast. Vinyl ketones were also reduced completely, while some other ketones were only partially reduced. An interesting study was recently completed by Idota and co-workers *(48)* who added 4 ppb of radioactive trans-2-nonenal to wort. Only 2% of the C-14 trans-2-nonenal was recovered from the fresh beer after fermentation. They did show that trans-2-nonenal added to beer can form a complex with excess sulphite (more than tenfold), but only 10% of the trans-2-nonenal was released from its complex even when the concentration of free sulphite was reduced to zero.

Importance Of The Redox State Of The Product

Oxidation-reduction reactions occur throughout the brewing process. They start after harvesting of the raw materials and continued up to the time when the finished product is consumed. Redox reactions that occur in the brewhouse are as important as those occurring in the final product *(31)*.

van Strien *(51)* describes an elegant method of measuring the redox coefficient rH which is a measure of the redox potential that is independent of pH. The rH scale ranges from zero (strongly reducing) to 42 (strongly oxidizing). A low redox coefficient means that the product is in a more reducing environment and is not as susceptible to oxidative changes. During fermentation, the reducing environment removes oxygen and reduces many of the carbonyls. At the end of fermentation rH is at its lowest point. van Gheluwe and co-workers *(50)* describe the changes in rH before and after fermentation by means of a spring analogy: if the rH is high prior to fermentation the spring will be compressed greatly during fermentation and has the potential to expand to a high rH level again after fermentation. To improve flavor stability, it is therefore important to keep the rH level during brewhouse operations as low as possible.

The primary reducing substances of natural origin, over which it is possible to exert control during processing, are the polyphenols, the melanoidins and the reductones *(8)*. When reducing sugars are strongly heated, primary decomposition products are formed which subsequently condense into slightly colored reductones *(52)*. In the presence of nitrogen compounds such as certain amino acids, the condensation products are strongly colored melanoidins, though some reductones may also be formed. The reaction is stronger in alkaline medium slowing down to about zero at a pH of 4.5. Melanoidins are formed mainly during kilning and reductones during wort boiling provided this is sufficiently long. Sugars added to the kettle should be added as early as possibly to promote reductone formation *(52)*. Brown *(46)* showed that darker malts such as crystal malt have high reducing power and beers produced with them produce less carbonyls which was confirmed by analytical and sensory analysis.

It should be noted, however, that melanoidins can be present in a reduced or oxidized state. Excessive agitation or turbulent transfer of hot wort will oxidize most of the melanoidins *(31)*. These oxidized melanoidins may then take part in the oxidation of higher alcohols to form stale carbonyls *(7)*.

Ohtsu and co-workers *(53)* measured changes in the level of the redox state by a variety of methods. They determined that when mash and wort absorbed a lot of oxygen during mashing and kettle boiling, the redox state of the resultant wort changed to the oxidized state and the flavor stability of the resulting product became poor. They suggest that polyphenols such as catechin and proanthocyanidins can act as antioxidants in wort and keep the wort in a more reduced state. The enzymatic oxidation of these polyphenols is increased when large amounts of oxygen are absorbed during mashing. The oxidative polymerization of these polyphenols and of melanoidins is enhanced during wort boiling when excessive oxygen is absorbed. Both oxidative reactions therefore decrease the levels of these natural antioxidants.

Irwin and co-workers *(41)* studied the effect of various polyphenols such as gallic acid, pyrogallol, delphinidin and myricetin act as pro-oxidants, whereas dihydroxy polyphenols such as monomeric and dimeric catechins function as antioxidants. Overexposure of wort to oxygen may lead to a loss of dihydroxy polyphenols and a resulting loss in flavor stability of the final product.

Importance of Oxygen

It is well know that high levels of air in the he final package leads to beer with a poorer flavor stability *(7, 14, 17, 56, 57)*. Whether the source of the problem are amino acids, higher alcohols, isohumulones, or unsaturated fatty acids, oxygen is considered the single most important negative factor *(49)*. Brewers have therefore attempted for many decades to minimize the introduction of oxygen into the product after fermentation and during packaging. During the last decade much more attention has also been given to reducing the amount of oxygen pickup in the brewhouse to maintain the wort in a more reduced state as discussed above.

The importance of oxygen can best be illustrated by means of the following two graphs in which the oxidation flavor that develops during storage is given for various levels of package oxygen. Note that total package oxygen here is measured indirectly;by measuring total package air by means of the so-called Zahm shake-out method. Better direct oxygen measurements are now available *(54, 55)*. The oxidation scores in Figures 1 and 2 are the mean scores obtained by a trained panel of about 30 tasters that are sensitive to the oxidation character that develops in this particular type of lager beer. Products are stored at 24°C and at 0°C for up to four months. The product stored at 0°C does not develop any noticeable oxidation flavor. Product that is stored at 24°C and subsequently transferred to 0°C maintains a given level of oxidation. For the test, products are removed bi-weekly from the 24°C storage room and placed in the 0°C room until all samples are collected after four months. The samples are then presented to the panel in a randomized fashion. Panelists are always given two samples for their evaluation: one stored at 24°C for a period of time and one stored only at 0°C; panelists are not told which sample is which.

The results in Figure 1 were obtained in 1979 when package air levels of 0.5 ml/12 oz. package were an excellent target for large breweries. The upper curve is for bottles with an initial level of 1.0 ml of air. This product oxidized significantly faster than that packaged in cans which had only 0.5 ml of air (lower curve). Two factors play a role here: 1) the initial air level in bottles was twice as high; and, 2) bottles are not completely air tight because of gas diffusion through the crown liner, so more oxygen enters the package during storage.

Figure 2 shows results obtained in 1992 with product in cans that had an even better air control. The upper curve is for cans with 0.30 ml of air/12 oz., while the lower curve is for 0.18 ml of air. In spite of the fact that air levels vary by almost a factor of two, it can be seen that the mean oxidation scores are relatively close. The results appear to indicate that even if package air could be brought down below 0.1 ml or even lower, perfect flavor stability would very likely not be obtained. Hashimoto *(7)* also concludes that decreasing package air reduces flavor

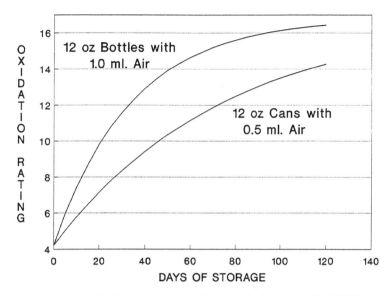

Figure 1. Oxidation flavor development in beer stored at 75°F.

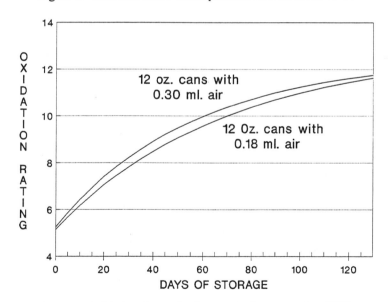

Figure 2. Oxidation flavor development in beer stored at 75°F.

staling, but does not prevent it altogether. Part of the reason for this is that the amount of molecular oxygen in the package is probably much less important than presence of active forms or oxygen, which in very small amounts can do a lot of damage. In fact, molecular oxygen may not be directly involved in the package in the formation of volatile carbonyls from their precursors.

Measurements of package oxygen as a function of time show that oxygen levels can decrease quite rapidly after filling depending on temperature. During pasteurization, a decrease in oxygen of about 15-25% was observed (56, 58). After pasteurization, the decrease in oxygen in cans appears to follow a first order reaction pattern (56). The original oxygen is reduced to less than 20% in ten days at 37°C or in six weeks at 4°C. What is interesting is that for beer stored for many months at temperatures of 4°C or below no oxidized flavor is noticeable even though most of the oxygen has been consumed. By introducing isotopic ^{18}O into the headspace of freshly bottled beer, Owades and Jakavac (59) were able to follow the fate of oxygen during storage. They found that 65% was incorporated in polyphenol molecules, 5% by isohumulones and 35% by volatile carbonyls. As discussed above, the uptake of oxygen by polyphenols will lower the reduction potential of the beer and make it more susceptible to oxidation.

Molecular oxygen may set the stage for oxidative degradation of volatile carbonyl precursors, but it is only involved in the oxidative reaction scheme through its free radical forms. Kaneda, et al (60) showed that hydrogen peroxide was generated during beer storage. The production of peroxide increased with increasing amount of headspace oxygen and with increasing temperature. Bamforth, et al (61) and Namiki (62) describe in detail the electron configuration and reactivities of various forms of oxygen and their involvement in product oxidation. Oxygen has two unpaired electrons each occupying a different orbit and having parallel spins. For electrons from other molecules to pair off, the acquired electrons must be of opposing spin than the O_2 electrons. This imposes major restrictions and as a consequence oxygen reacts very slowly with many non-radical species.

If oxygen acquires an electron, then superoxide ($O_2 \overset{\circ}{-}$) is produced. Since this molecule only has a single electron to pair off, it is more reactive. Acquisition of a further electron yields peroxide O_2^{2-}. Peroxide such as hydrogen peroxide decompose rapidly under the influence of energy (light) yielding hydroxyl molecules (2 OH°) or singlet oxygen $^1O_2$. Hydroxyl molecules are very reactive with extremely high rate constants for reactions with biological materials such as fatty acids, sugars, alcohols, and polyphenols. The conversion of superoxide or peroxide to hydroxyl is accomplished through the interaction of metal catalysts such as Cu^+ or Fe^{2+} involving the Fenton or Haber-Weiss reactions. The Haber-Weiss reaction shown below was determined (41) to be more favorable thermodynamically than the Fenton reaction.

$$M^{n+} + H_2O_2 + H^+ \Rightarrow M^{(n+1)+} + H_2O + OH^\circ$$

$$M^{(n+1)} + O_2^{\circ-} \Rightarrow M^{n+} + O_2$$

$$\text{Net:} \quad O_2^{\circ-} + H_2O_2 + H^+ \Rightarrow H_2O + O_2 + OH^\circ$$

In this reaction, M^{n+} is Fe^{2+} or Cu^+, but Cu^+ is more reactive than Fe^{2+} by orders of magnitude.

The importance of active oxygens in beer flavor deterioration was illustrated by Kaneda and co-workers (63). They added a compound CLA to beer which only reacts with superoxide and with singlet oxygen and shows a marked luminescence on doing so. A good correlation was found between the generation of luminescence as result of active oxygens and the degree of staling determined by a sensory panel. Different brands and types of beers produced different amounts of luminescence and different amounts of stale flavor development (64). Luminescence was also accelerated and higher intensities were reached with increasing storage temperatures. It was concluded that superoxide is actively involved in beer staling since superoxide scavengers such as SO_2 and ascorbic acid were found to reduce luminescence as well a stale flavor development. It was suggested that deterioration rates of beers may be assessed from the chemiluminescence producing patterns in the fresh beers before storage.

Control of Raw Materials and Brewhouse Operations

Barley, Malt and Adjuncts Even though the concentration of trans-2-nonenal in dark malts is about twice that of pale malts (34), most investigators believe that beers brewed with dark malt are inherently more stable than beers brewed with pale malts. There are several reasons for this. First of all, almost all trans-2-nonenal in malt is lost during brewhouse operations and of the small amount that is carried on to fermentation only 2% is found in the final beer (48). Secondly, the higher kilning temperatures used in the preparation of dark malts inactivate all fatty acid degrading enzymes, so there is no chance for enzymatic precursor formation in the brewhouse. Thirdly, dark malt contains a high level of melanoidins which have reducing capacity, unless they are over-oxidized in the brewhouse. Drost, et al (18) showed that when higher kilning temperatures were used (as in dark malt production), the final product had a low nonenal potential.

Since activities of lipoxygenase and hydroperoxide isomerase are dependent on barley variety and growth location, raw material selection may be optimized. It seems also important to control microflora growth during malting since microbially produced fatty acid degrading enzymes are apparently more stable.

The effect of adjuncts on flavor stability was studied by Peppard, et al *(65)* on a pilot scale with 30% substitution of malt by adjuncts. Rice grits and corn grits decreased flavor stability compared to all malt, while the use of wheat flour, barley grits, and corn syrup improved the flavor stability.

Oxygen Control in the Brewhouse The main reason for minimizing oxygen pickup in the brewhouse is to maintain the wort in reduced state *(53)*. When mashing was done under a nitrogen atmosphere, Brown *(46)* showed that the reducing power of the resulting mash remained higher and the flavor stability increased. A further improvement was noted when milling was also carried out under nitrogen. In actual brewing practice, maintaining a nitrogen atmosphere in the brewhouse may not be practical, but it does point out the importance of minimizing oxygen pickup.

Narziss and co-workers *(66, 14)* did extensive laboratory and commercial scale tests in which brews were purposely aerated during mashing and/or lautering or nitrogen gassing was applied (laboratory only). Results from the pilot plant clearly show the negative influence of oxygen on flavor stability. This negative influence was also found in tests done in one commercial brewery. Tests in two other breweries showed no significant effects, as other factors apparently were more important.

Currie, et al *(73)* found more highly oxidized pilot plant worts to develop significantly more oxidized beers after five weeks of storage at 30°C.

The effect of brewhouse aeration and nitrogen gassing on flavor stability is apparently hard to determine in practice as Hug and co-workers *(67)* and Schur *(68)* could not find any significant influence in experiments done on a large pilot scale and on a commercial scale respectively. However, their taste evaluations were done after four or five weeks of storage of the final product at room temperature and with a relatively small panel.

The amount of oxygen absorption by the mash or wort during various brewhouse operations may be determined by means of the sulphite method developed by Lie, et al *(69)*. The following is a summary of steps that can be taken to reduce oxygen absorption suggested by Lie *(69)*, Stippler *(70)*, and by Zürcher and Grass *(71)*:

- Use a premasher to thoroughly mix the grist with the mash or cooker water. A premasher allows for thicker mashes and avoids the use of high agitation speeds required for wetting the grains. The use of a specially designed doughing worm as premasher was found helpful *(71)*.

- Use low agitation speeds where possible and only mix when the agitator is fully covered.

- Enter all brewhouse vessels from below and avoid splashing and wort spreaders to distribute the mash in the lauter tun.

- Avoid air pickup during mash and wort pumping as a result of leaky pipe connections.

- Keep spent grain below the wort surface.

- Avoid air swirls during emptying of the underback.

- During brew kettle boil, avoid air from sweeping across the surface of the wort, removing the protective layer of foam and steam.

Brewhouse Variables and Materials of Construction Drost, et al *(16)* showed that when malt was mashed in at 67°C the resulting beer had a lower nonenal potential and a better flavor stability. This result suggests that the enzymes involved in the production of nonenal precursors from fatty acids are largely inactivated at this temperature. This method is not practical, however, since most saccharifying enzymes will be inactivated as well.

A more practical method may be to reduce the pH during mashing which also lowers the nonenal potential *(16)* and produces a beer with improved flavor stability. Narziss *(72)* suggests reducing the pH from 5.5 to a level of about 5.1.

Since the transition metals such as Fe^{2+} and Cu^+ are able to catalyze various enzymatic and non-enzymatic oxidation reactions in the brewhouse, it is preferred to use stainless steel as the material of construction rather than copper, which was the material traditionally used to obtain good heat transfer. van Gheluwe *(58)* as well as Narziss *(66)* noted the poorer flavor stability of the product when using copper.

The length of kettle boil and the duration of the hot wort stand *(72)* appear to be significant parameters affecting flavor stability. The percent evaporation was found to be unimportant *(74)* as long as at least 2% was evaporated. The longer wort components are exposed to high temperatures in the presence of oxygen and metal catalysts, the more free radical formation and auto-oxidation can take place. Buckee and Barrett *(74)* found that a kettle boil time of one hour gives a more flavor stable beer than a time of only 0.5 hour. This result appears to contradict the findings of Ohtsu *(53)* who showed that the reducing power of wort declines continuously during kettle boil.

Narziss *(72)* suggested to cool knock-out wort to 80°C on the way to the hot wort settling tank in order to improve flavor stability.

Control of Lipids in the Wort Even though it is widely thought that lipids are the main source of trans-2-nonenal in the final product, there is not much experimental evidence that reducing the amount of lipids or fatty acids in the malt or wort decreases the cardboard flavor development caused by trans-2-nonenal.

Currie, et al *(73)* reduced the lipid level in malt by means of CO_2 extraction from 3% to 0.5% and found no significant change in oxidized flavor development. A similar experiment was done by Peppard, et al *(65)* with the same results. Could it be possible that oxidative lipid degradation products that are formed during malting are not efficiently extracted with CO_2 ?

Lipids, including oxidized lipids are preferentially absorbed by spent grains and wort sediments. A good correlation between lipids and suspended solids was obtained by Whitear, et al *(75)*. Drost and co-workers *(16)* showed that when suspended particles are carried over into the pitching wort by poor separation techniques, products with poor flavor stability result. The same was shown by Narziss *(14)* who obtained an improvement in flavor stability when the turbid wort from a mash filter was clarified by means of diatomaceous earth filtration.

According to Whitear and co-workers *(75)*, the presence of high levels of lipids in wort going to the brew kettle has much less of an effect on flavor stability than the effect of oxygen pickup during processing. Beers prepared from worts to which an additional 190 mg/L of lipids were added from spent grains actually had a slightly better flavor stability than their control. What is important for flavor stability according to these investigators as well as to Olsen *(76)* is the amount of lipids carried over to the fermenter with trub. Especially the carry-over to fermentation of spent grain particles with a sieve fraction of less than 0.25 mm had a detrimental effect on flavor stability. Lipids extracted from these fatty acids show the same effect. Note that carry-over of lipids or fatty acids to fermentation decreases ester formation and increases the formation of higher alcohols *(76)*.

Wort turbidity can be controlled by the following measures:

- The choice of wort separation equipment: deep bed mash tun is the best, followed by lauter tun, strain master, and mash filter *(75)*. New generations of mash filters are better *(77)*.

- Longer initial recycle in wort separation.

- Fewer mash-ups in lauter tuns, and a proper recycle after mash-ups.

- Use of Whirlpools and other efficient trub separation equipment.

- Use of flocculating agents to increase the settling efficiency of the fine suspended materials which appear to be the most detrimental.

- Use of cold wort filtration prior to fermentation.

Control of Cellar Operations

Fermentation Control Kaneda and co-workers *(78)* have shown that fermentation conditions can have a definite effect on the flavor stability of the beer. The flavor stability increased with a decrease in fermentation temperature. This increased flavor stability coincided with a decrease in chemiluminescence and with an increase in SO_2 produced during fermentation. As pointed out before, SO_2 apparently has a flavor stabilizing effect in packaged product through its ability to complex with volatile aldehydes.

Most investigators believe that SO_2 stabilizes the product flavor whether this SO_2 is produced during fermentation or added as an antioxidant later in the process, although this is questioned by Dufour *(79)*. He argues that especially toward the end of fermentation, SO_2 complexes with carbonyls such as diacetyl and acetaldehyde which are no longer reduced quickly enough by yeast. These complexed carbonyls are then carried into the product. Nordlöv *(80)* determined that at the end of fermentation acetaldehyde and SO_2 are present at equimolar concentrations.

SO_2 levels at the end of fermentation are closely related to yeast growth: the slower the yeast growth the higher the SO_2. Generally conditions with a high SO_2 correspond to an high pH. Beers with a high pH are usually more flavor stable and taste less oxidized than beers with low pH *(81)*. It is hard to determine whether it is the higher pH or the higher SO_2 that is most beneficial for flavor stability. SO_2 is formed by yeast as an intermediate in the sulphur amino acid biosynthetic pathway and is produced in excess of the requirements by some yeasts in certain circumstances and excreted into the fermenting beer *(82)*.

Methods to produce high final SO_2 levels are:

- Low fermentation temperatures *(78)*.

- Low degree of wort aeration *(80)*.

- Low lipid carry-over from the brewhouse *(83)*.

- Low yeast pitch rate *(80)*.

- Low original gravity of the wort *(80)*.

- High sulphate level in the wort *(87)*.

Oxygen Control Once fermentation is completed and yeast is removed from the beer, the product is again susceptible to oxidation and care must be taken to avoid or minimize all contact with oxygen. A good way to monitor proper beer handling practices is to measure dissolved oxygen levels at various strategic points in the cellar operations as discussed by Klimovitz *(84)*. The practical control of oxygen in the cellars was also discussed by Thomson *(85)*.

The following is a summary of oxygen control procedures:

- Use CO_2 , or nitrogen, or deaerated water to move product between storage tanks or when first filling a filter.

- Use tight couplings to prevent air leaks on the suction side of pumps.

- Deaerate filter aid slurries by means of CO_2 or N_2.

- Blanket tanks with CO_2 or N_2 before filling them with product.

- Use CO_2 or N_2 counterpressure on all tanks.

- Use vortex breakers or baffles during emptying tanks.

- Only use CO_2 with low (less than 10 ppm) oxygen levels *(86)*.

- Deoxygenate alcohol adjustment water to less than 30-40 ppb.

If the above procedures are properly used, beer going to packaging can have dissolved oxygen levels not exceeding 50-70 ppb.

Chillproofing Materials Adsorbents such as bentonite, PVPP, and silica hydrogel are commonly used to improve the physical stability of beer. These materials have also been studied as to their effect on flavor stability.

Peppard and co-workers *(65)* determined that silica hydrogel consistently improved flavor life, whereas the others did not. Beers treated with silica hydrogel were found to have a greater reducing potential after treatment. Analysis of adsorbed materials showed numerous higher alcohols and fatty acids, including trihydroxy linoleic acid.

Beers chillproofed with PVPP developed significantly higher levels of typical oxidized flavors than beer chillproofed with silica hydrogel *(6)*. This may either be due to the beneficial effect of silica hydrogel or due to the adverse effect of PVPP as a result of polyphenol adsorption.

Narziss and co-workers *(66)* added PVPP to wort in the brew kettle and showed considerably reduced levels of certain polyphenols. Even fresh, the resulting beer appeared oxidized, and it became strongly oxidized after one week at 27°C. This suggests that a certain quality of polyphenols is required to prevent excessive oxidation of other components in the brewhouse.

Antioxidants and Oxygen Scavengers

Since beer flavor instability is caused by oxidative degradation of beer compounds, brewers may choose to augment the endogenous antioxidants such as reductones, phenolic compounds and sulphite with exogenous antioxidants to improve flavor stability. Excellent general reviews of the use of antioxidants in food systems are given by Namiki *(62)* and by Shahidi and Wanasundara *(88)*. Oxygen is of prime importance in oxidation reactions either through direct or indirect involvement. Systems have therefore been developed to scavenge oxygen either during processing or in the package.

Antioxidants are compounds that are capable of delaying, retarding, or preventing auto-oxidation processes. In the auto-oxidation of fatty acids, for example, an antioxidant will donate hydrogen to the fatty acid radical. The resulting antioxidant radical must be much more stable than the fatty acid radical to be effective. Antioxidants may also inactive oxygen radicals by donating hydrogen or electrons.

Antioxidants, when added to beer, may result in off-flavors. SO_2 at high levels causes a burnt match aroma and flavor. Isoascorbic acid was found to produce an undesirable off-flavor in canned beer, possibly by reacting with components in the can lining binder *(89)*. Off-flavors with ascorbates or isoascorbates can also develop as a result of the formation of dehydroascorbates which actively participate in browning reactions *(90)*. These problems do not arise when sulphites are also added since SO_2 keeps isoascorbate in an inactive reduced form.

SO_2 and isoascorbates are relatively poor scavengers of molecular oxygen and they were found to have little effect on the rate of oxygen uptake in beer *(58, 90)*.

However, they are much better scavengers of active oxygen, especially superoxide. By trapping the active oxygens, they inhibit the free radical reactions during the storage of beer resulting in improved flavor stability (63). A combination of potassium meta bisulphite and sodium isoascorbate (40:60) was found to be significantly more effective than using sodium isoascorbate by itself (87). Vilpola (91) also found that SO_2 and ascorbic acid work best when they are used together. The additive effect is traced to their different non-competitive anti-oxidative mechanisms.

Molecular oxygen may be scavenged by various immobilized enzyme systems. A glucose oxidase/catalase system, for example, may be incorporated in a can liner (89), or in a crown liner (93), or the enzymes may be crosslinked with gluteraldehyde to a carrier and used in a plate and frame filtration system (92). Glucose is needed as a substrate for this system and gluconic acid is formed as a by-product. Various other oxidase systems may also be used such as oxalate oxidase, lactate oxidase and amino acid oxidase (94).

Blockmans and co-workers (95) determined that the glucose oxidase, catalase, glucose system added directly to beer was very efficient in removing dissolved oxygen, but is was totally inefficient in stabilizing flavor as the resulting products were highly oxidized. They suggested that catalase may not have been sufficiently effective in the reduction of peroxide to water and half of the original dissolved oxygen, but that catalase may have promoted the reaction of peroxide with other oxidizable substances such as alcohols. When SO_2 was added in combination with the other ingredients, flavor stability was improved, probably because the sulphite is preferentially oxidized compared to alcohols, although SO_2 is also effective in complexing volatile aldehydes, as discussed above.

A proven oxygen scavenger is also yeast (97). Naturally conditioned beers with yeast in the bottle can remain drinkable for years, but require skill for proper handling in the brewery and in the trade (4). At higher storage temperatures, yeast can autolyze and produce sulphidic off-flavors in the product. It has even been found that when stale beer is treated with a mixture of active yeast and glucose, the stale character is greatly reduced and papery, musty flavor is eliminated (65).

A potentially very effective way of improving flavor stability is by using enzyme systems that scavenge activated forms of oxygen. This can be done, for example, by adding superoxide dismutase (SOD) and catalase to the fermenter through the addition of soy extract which is high in SOD (98). SOD catalyses the reduction of superoxide to peroxide, which is reduced under the influence of catalase to water and molecular oxygen. Superoxide dismutase and catalase are also present in barley and malt and increase during malting. They are completely inactivated during mashing (99). It is essential that transition metal ions are kept to a minimum as they will rapidly convert superoxide and peroxide to the very reactive hydroxyl. Copper ions are the most potent activating agent, but they can be rendered inactive by chelation. Iron can be active even in a chelated form. In this respect, the addition of ascorbic acid is helpful, as it is able to chelate heavy metal ions *(62)* as well as scavenge hydroxyl.

Control of Packaging Operations and Materials

If care has been taken during cellar operations to minimize oxygen introduction into the product, dissolved oxygen levels in beer going to packaging should not exceed 50 to 70 ppb. During filling and closing total package oxygen levels (dissolved oxygen plus headspace oxygen) will increase several fold to levels of about 200 to 350 ppb (equivalent to about 0.2 to 0.35 ml air/12 oz.). Note that only two decades ago total package oxygen concentrations of over 1000 ppb were common. Technological advances have therefore made great strides in minimizing oxygen pickup, but further advances are still possible through optimization. Peterson and Evans *(101)* identified 97 control factors that could influence package oxygen. They optimized the 10 most important ones by means of a Taguchi method.

The following general procedures were found to be most important:

- Avoid oxygen pickup during transport of beer to the filler bowl.

- Counterpressure the filler bowl with inert gases, such as CO_2 or nitrogen. Continuously purge the filler bowl headspace with these gases to prevent a buildup of oxygen, which enters the bowl as it is being displaced with beer in the containers.

- Pre-evacuate containers once or twice *(100)* and apply fresh CO_2 to the containers between both evacuations (less CO_2 is then required for bowl purge). Proper maintenance is vacuum pumps is essential.

- Use high pressure water jets to nucleate CO_2 in the beer in filled bottles *(102)*. This CO_2 strips dissolved oxygen from the beer and it creates a fine foam which displaces headspace air from the bottle.

- For cans, use bubble breakers and rail gassing. Bubble breakers work by impinging CO_2 on top of the beer to break the larger foam bubbles which often contain higher concentrations of air. Rail gassers work by providing a blanket of CO_2 or nitrogen over the beer on the transfer between the filler and closer *(103)*.

- Use properly positioned under-cover gassers prior to seaming the lids on cans.

During the mid-1980's several groups of investigators *(104,105)* discovered that bottle crowns do not provide a hermitic seal and that oxygen is able to permeate through the crown liner during its entire storage life. Only a negligible amount of oxygen apparently diffuses through can seals.

The permeation rate in bottles depends on the composition and shape of the crown liner material and on the temperature *(106)*. At room temperature, permeation rates vary from 0.001 to 0.002 ml O_2/bottle/day. For a 12 oz. bottle and a time span of 100 days, the total oxygen ingress is equivalent to 400 to 800 ppb. This is two to four times as much as the amount of oxygen that is present right after filling.

Development work has focused on materials that reduce or eliminate oxygen permeation through crown liners. A crown cork with an aluminum spot inlay virtually eliminated oxygen ingress *(104)*, while novel polymers cut the ingress in half *(106)*, but these materials are not considered economical.

Various oxygen scavenging materials were also developed for incorporation in or on the crown liner. These scavengers are not only able to eliminate oxygen that comes in through the liner, but also to reduce oxygen that is present in the headspace of the bottle after filling. However, the scavenging action is not fast enough to prevent oxygen from reacting with beer components during pasteurization.

Oxygen scavenging systems include:

- Glucose oxidase/catalase and its substrate glucose, entrapped in a microporous gel layer on the crown *(93)*.

- Active dry yeast, embedded in a moisture permeable polymer attached to the crown *(107)*. This yeast is able to withstand pasteurization temperatures.

- Ascorbates or isoascorbates, along or in combination with sulfites, in a polymeric matrix *(108)*.

- Ascorbates or isoascorbates dispersed throughout a polymer carrier along with a transition metal to catalyze the oxygen scavenging properties *(109)*. The transition metals may be salts or chelates with materials such as EDTA. Transition metal chelates also posses oxygen scavenging properties which will augment the oxygen scavenging capability of the ascorbates.

Control of Beer Storage and Handling Conditions

External influences that can effect beer during storage are temperature, light, motion, and time. There is only limited control by the brewery over these parameters.

Temperature/Time These are by far the most important of these factors. Hashimoto (7) noted that a distinct stale flavor develops in lager beer after 12 to 30 days at 30°C. At 8°C, it takes about 40-55 days to develop this flavor, whereas at 2°C only minimal change occur. Nordlöv and Winell (44) showed that the development of trans-2-nonenal and of a typical cardboard flavor is a factor 6 to 7 greater at 30°C than it is at 20°C.

The importance of temperature is clearly demonstrated in Figure 3 which shows the rating of the oxidation character that develops in a typical American lager over time at various temperatures. By comparing the average panel ratings in this figure to those in Figures 1 and 2, it can be seen that the harmful effect of a high storage temperature on stale flavor development is much greater than the effect of package oxygen, a conclusion that was also reached by Hashimoto (7).

The effects can be minimized by encouraging temperature control or refrigeration at distributor and retail levels, rapid turnover, inventory rotation, and the proper use of good-before-dates (29).

Light When beer is exposed to sunlight, a pronounced sulphury flavor can develop. This unpleasant, skunk-like, flavor is usually termed "sunstruck flavor" and may be a problem for beers packaged in clear or green bottles (2). This flavor can also develop after beer is poured into a glass. Sunstruck flavor is attributed to the formation of 3-methyl-2-butene-1-thiol by photolysis of iso-alpha-acids in the presence of sulphur containing amino acids, but it may also be due to a light-induced formation of methyl-mercaptan and hydrogen sulfide. This problem can be minimized by protecting the beer from long wave UV light by using appropriate packaging materials or by using a light-stable hop product (16).

Various oxidation reactions, that were discussed in the section on mechanisms of beer oxidation, may also be catalyzed by light. Devreux and co-workers (1) showed that a Strecker degradation of amino acids is more rapid in light and that it is catalyzed by riboflavin present in beer. The same is true for the oxidative degradation of isohumulones. The oxidation of alcohols by melanoidins is very fast in the he presence of light and riboflavin, but it is quickly inhibited by small amount of polyphenols. Light also catalyzes the auto-degradation of fatty acids, but this reaction is inhibited by riboflavin. Whichever mechanism is important for stale flavor development, the exclusion of light is likely to be beneficial.

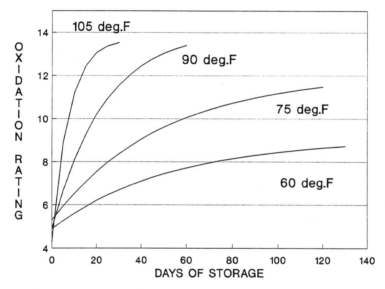

Figure 3. Oxidation flavor development in beer with 0.3 ml air/12 oz.

Motion The effect of motion during transportation on flavor stability was investigated by Miedaner, et al *(3)* by means of a model study. They found that the development of oxidized flavor during storage at 40°C for one week was significantly increased when the high temperature storage was preceded by one day of shaking. Mixing enhances the diffusion of headspace oxygen into the beer, so that reactions of oxygen with beer can proceed at a higher rate.

In practice, it is hard to reduce the effects of this factor, unless perhaps beer is packaged in larger containers such as kegs in which the headspace volume is small compared to the beer volume.

Acknowledgment

The author would like to thank the management of Miller Brewing Company for permission to publish this review.

Literature Cited

1. Devreux, A.; Blockmans, C; van de Meerssche, J. *EBC Monograph VII, EBC Flavor Symposium* **1981**, pp. 191-201.
2. Laws, D.R.; Peppard, T.L. *Food Chemistry* **1982**, *vol 9*, pp. 131-146.
3. Miedaner, H.; Narziss, L.; Eichhorn, P. *Proceedings EBC Congress* **1991**, pp. 401-408.
4. Dalgliesh, C.E. *Proceedings EBC Congress* **1977**, pp. 623-659.
5. Whitear, A.L. *EBC Monograph VII, EBC Flavor Symposium* **1981**, pp. 203-210.
6. Whitear, A.L.; Carr, B.L.; Crabb, D.; Jackques, D. *Proceedings EBC Congress* **1979**, pp. 13-25.
7. Hashimoto, N. In *Brewing Science*; Pollock, J.R., Ed.; Academis Press; *vol. 2*; pp. 347-405.
8. Chapon, L. In *Brewing Science*; Pollock, J.R., Ed.; Academic Press; *vol. 2*, pp. 407-456.
9. Wheeler, R.E.; Pragnell, M.J.; Pierce, J.S. *Proceedings EBC Congress* **1971**, pp. 423-436.
10. Hashimoto, N. *Journal Inst. of Brewing* **1972**, *vol. 78*, pp. 43-51.
11. Meilgaard, M.C. *MBAA Technical Quarterly* **1975**, *vol. 12*, pp. 107-117.
12. Meilgaard, M.C. *MBAA Technical Quarterly* **1975**, *vol. 12*, pp. 151-168.
13. Greenhoff, K.; Wheeler, R.E. *Journal Inst. of Brewing* **1981**, *vol. 86*, pp. 35-41.

14.   Narziss, L. *Journal Inst. of Brewing* **1986**, *vol. 92*, pp. 346-353.
15.   Meilgaard, M.C. *MBAA Technical Quarterly* **1974**, *vol. 11*, pp. 117-120.
16.   Drost, B.W.; van den Berg, R.; Freijee, F.J.; van der Velde, E.G.; Hollemans, M. *Journal Am. Soc. Brew. Chem.* **1990**, *vol. 48*, pp. 124-131.
17   Wang, P.S.; Siebert, K.J. *MBAA Technical Quarterly* **1974**, *vol. 11*, pp. 110-117.
18.   Wang, P.S.; Siebert, K.J. *Journal Am. Soc. Brew. Chem.* **1974**, *vol. 32*, pp. 47-49.
19.   Jamieson, A.M.; van Gheluwe, J.E. *Am. Soc. Brew. Chem. Proc.* **1970**, *vol. 28*, pp. 192-197.
20.   Visser, M.K.; Lindsay, R.C. *MBAA Technical Quarterly* **1971**, *vol. 8*, pp. 123-128.
21.   Drost, B.W.; van Eerde, P.; Hoekstra, S.F.; Strating, J. *Proceedings EBC Congress* **1971**, pp. 451-458.
22.   Schmitt, D.J.; Hoff, J.F. *Journal Food Sci.* **1979**, *vol. 44*, pp. 901-904.
23.   Eichhorn, P.; Komori, T.; Miedaner, H.; Narziss, L. *Proceedings EBC Congress* **1989**, pp. 717-724.
24.   Macias, J.R.; Espinosa, I. *Beverages* **1981**, pp. 14-17.
25.   Blockmans, C.; Devreux, A.; Masschelein, C.A. *Proceedings EBC Congress* **1975**, pp. 699-713.
26.   Hashimoto, N.; Kuroiwa, Y. *Journal Am. Soc. Brew. Chem.* **1975**, *vol. 33*, pp. 104-111.
27.   Barker, R.L.; Gracey, D.E.; Irwin, A.J.; Pipasts, P.; Leiska, E. *Journal Inst. of Brewing* **1983**, *vol. 89*, pp. 411-415.
28.   Hashimoto, N. *Journal Inst. of Brewing* **1972**, *vol. 78*, pp. 43-51.
29.   Hashimoto, N. *Rept. Res. Lab. Kirin Brewery Co.* **1988**, *vol. 31*, pp. 19-32.
30.   Hashimoto, N.; Eshima, T. *Journal Am. Soc. Brew. Chem.* **1977**, *vol. 35*, pp. 145-150.
31.   Fix, G.J. *Principles of Brewing Science*; Brewers Publications, **1990**.
32.   Hashimoto, N.; Shimazu, T.; Eshima, T. *Rept. Res. Lab. Kirin Brewery Co.* **1979**, *vol. 22*, pp. 1-10.
33.   Yabuuchi, S. *Agr. Biol. Chem.* **1976**, *vol. 40*, p. 1987.
34.   Tressl, R.; Bahri, D.; Silwar, R. *Proceedings EBC Congress* **1979**, pp. 27-41.
35.   van Eerde, P.; Strating, J. *EBC Monograph VII, EBC Flavor Symposium* **1981**, pp. 117-121.
36.   Stenroos, L.; Wang, P.; Siebert, K.; Meilgaard, M. *MBAA Technical Quarterly* **1976**, *vol. 13*, pp. 227-232.

37. Dominguez, X.A.; Canales, A.M. *Brewers Digest* **1974, July,** pp. 40-47.
38. Garza Ulloa, H.; Villareal, R.; Canales, A.M. *Brewers Digest 1976,* **April,** pp. 48-40, 72.
39. Baxter, E.D. *Journal Inst. of Brewing* **1982,** *vol. 88,* pp. 390-396.
40. Schwarz, P.B.; Pyler, R.E. *Journal Am. Soc. Brew. Chem.* **1984,** *vol. 42,* pp. 47-53.
41. Irwin. A.J.; Barker, R.L.; Pipasts, P. *Journal Am. Soc. Brew. Chem.* **1991,** *vol. 49,* pp. 140-149.
42. Kaneda, H.; Kano, Y.; Kamimura, M.; Osawa, T.; Kawakishi, S. *J. Agric. Food Chem.* **1990,** *vol. 38,* pp. 1363-1367.
43. Clarkson, S.P.; Large, P.J.; Hegarty, P.K.; Bamforth, C.W. *Proceeding EBC Congress* **1989,** pp. 267-274.
44. Nordlöv, H.; Winell, B. *Proceedings EBC Congress* **1983,** pp. 271-278.
45. Gracey, D.E.F.; Barker, R.L.; Irwin, A.J.; Pipasts, P.; Leiska, E. *Inst. of Brew., Austr. and N.Z. Section* **1984,** *vol. 18,* pp. 50-58.
46. Brown, J.W. *Ferment* **1984,** *vol. 2, no. 1,* pp. 51-53.
47. Peppard, T.L.; Halsey, S.A. *Journal Inst. of Brewing* **1981,** *vol. 87,* pp. 386-390.
48. Idota, Y.; Shindou, K.; Arima, M.; Kamiya, T.; Mawatari, M. *Brewing Congress of the Americas* **1992,** St. Louis: to be published.
49. Bamforth, C. *The Brewer* **1986, February,** pp. 48-51.
50. van Gheluwe, G.E.A.; Valyi, Z. *MBAA Technical Quarterly* **1974,** *vol. 11,* pp. 184-192.
51. van Strien, J. *Journal Am. Soc. Brew. Chem.* **1987,** *vol. 45,* pp. 77-79.
52. Thomson, R.M. *Brewers' Guild Journal* **1952,** *vol. 38,* pp. 167-185.
53. Ohtsu, K.; Hashimoto, N.; Inoue, K.; Miyaki, S. *Rep. Res. Kirin Brew. Co.* **1983,** *vol. 26,* pp. 7-14.
54. Hale, J. *The Brewer* **1987,** pp. 363-365.
55. Jandrau, G.L.; Hahn, C.W. *Journal Am. Soc. Brew. Chem.* **1978,** *vol. 36,* pp. 44-52.
56. Hoag, L.,; Reinke, H. *Am. Soc. Brew. Chem. Proc.* **1960,** *vol. 18,* pp. 141-145.
57. Schmidt, F.; Rosendal, I.; Sejersen, L. *EBC Monograph VII, EBC Flavor Symposium* **1981,** pp. 103-115.
58. van Gheluwe, G.E.; Jamieson, A.M.; Valyi, Z. *MBAA Technical Quarterly* **1970,** *vol. 7,* pp. 158-166.
59. Owades, J.L.; Jakavac, J. *Am. Soc. Brew. Chem Proc.* **1966,** *vol. 24,* p. 180.
60. Kaneda, H.; Kano, Y.; Osawa, T.; Kawakishi, S.; Kamada, K. *Journal Am. Soc. Brew. Chem.* **1989,** *vol. 47,* pp. 49-53.

61. Bamforth, C.W.; Boulton, C.A.; Clarkson, S.P.; Large, P.J. *Inst. of Brew., Austr. and N.Z. Section, Proc. of 20th Conv.* **1988**, pp. 211-219.

62. Namiki, M. *CRC Crit. Rev. Food Sci. and Nutr.* **1990**, *no. 4*, pp. 273-300.

63. Kaneda, H.; Kano, Y.; Osawa, T.; Kawakishi, S.; Koshino, S. *Proceedings EBC Congress* **1991**, pp. 433-440.

64. Kaneda, H.; Kano, Y.; Kamimura, M. *Journal Inst. of Brewing* **1991**, *vol. 97*, pp. 105-109.

65. Peppard, T.L.; Buckee, G.K.; Halsey, S.A. *Proceedings EBC Congress* **1983**, pp. 549-556.

66. Narziss, L.; Reicheneder, E.; Lustig, S. *Brauwelt International* **1989**, pp. 238-250, and: **1990**, pp. 39-46.

67. Hug, H.; Anderegg, P.; Mändli, H.; Pfenninger, H. *Brauerei Rundschau* **1986**, *vol. 97*, pp. 113-116.

68. Schur, F. *Brauerei Rundschau* **1986**, *vol. 97*, pp. 117-120.

69. Lie, S.; Grindem, T.; Jacobsen, T. *Proceedings EBC Congress* **1977**, pp. 235-243.

70. Stippler, K. *MBAA Technical Quarterly* **1988**, *vol. 25*, pp. 54-61.

71. Zürcher, Ch.; Gruss, R. *Brauwelt International* **1989**, pp. 350-355.

72. Narziss, L. Award of Merit Lecture at *Brewing Congress of the Americas* **1992**, to be published in MBAA Technical Quarterly.

73. Currie, B.R.; Kulandal, J.; Fitzroy, M.D.; Hawthorne, D.B.; Kavanagh, T.E. *Inst. of Brew., Austr. and N.Z. Section, Proc. of 21st Conv.* **1990**, pp. 117-125.

74. Buckee, G.K. and Barrett, J. *Journal Inst. of Brewing* **1982**, *vol. 88*, pp. 329-331.

75. Whitear, A.L.; Maule, D.R.; Sharpe, F.R. *Proceedings EBC Congress* **1983**, pp. 81-88.

76. Olsen, A. *EBC Monograph VII, EBC Flavor Symposium* **1981**, pp. 223-236.

77. vanWaesberghe, J.W.M. *MBAA Technical Quarterly* **1991**, *vol. 28*, pp. 33-37.

78. Kaneda, H.; Kimura, T.; Kano, Y.; Koshino, S.; Osawa, T.; Kawakishi, S. *Journal Ferm. and Bioeng.* **1991**, *vol. 72*, pp. 26-30.

79. Dufour, J.P. *Proceedings EBC Congress* **1991**, pp. 209-216.

80. Nordlöv, H. *Proceedings EBC Congress* **1985**, pp. 291-298.

81. Grigsby, J.S.; Palamand, S.R.; Davis, D.P.; Hardwick, W.A. *Am. Soc. Brew. Chem. Proceedings* **1972**, *vol. 30*, pp. 87-92.

82. Brewer, J.D.; Fenton, M.S. *Inst. of Brew. Austr. and N.Z. Section, Proc. of 16th Conv.* **1980**, pp. 155-162.

83. Dufour, J.P.; Carpenter, B.; Kulakumba, M.; van Haecht, J.L.; Devreux, A. *Proceedings EBC Congress* **1989**, pp. 331-338.

84. Klimovitz, R.J. *MBAA Technical Quarterly* **1972**, *vol. 9*, pp. 63-68.

85. Thomson, R.M. *Brewers Guild Journal* **1952**, *vol. 38*, pp. 167-184.

86. Huige, N.J.; Charter, W.M.; Wendt, K.W. *MBAA Technical Quarterly* **1985**, *vol. 22*, pp. 92-98.

87.   Klimovitz, R.J.; Kindraka, J.A. *MBAA Technical Quarterly* **1989**, *vol. 26*, pp. 70-74.

88.   Shahidi, F.; Wansundara, P.K.J.P.D. *CRC Crit. Rev. Food Sci. and Nutr.* **1992**, *vol. 32*, pp. 67-103.

89.   Reinke, H.G.; Hoag, L.E.; Kincaid, C.M. *Am. Soc. Brew. Chem. Proceedings* **1963**, *vol. 21*, pp. 175-180.

90.   Brenner, M.W.; Stern, H. *MBAA Technical Quarterly* **1970**, *vol. 7.*, pp. 150-157.

91.   Vilpola, A. *Mallas Olut* **1985**, pp. 178-184.

92.   Hartmeier, W.; Willox. I.C. *MBAA Technical Quarterly* **1981**, *vol. 18*, pp. 145-149.

93.   Goossens, E.; Dillemans, M.; Masschelein, C.A. *Proceedings EBC Congress* **1989**, pp. 625-632.

94.   Prieels, J.P.; Heilporn, M.; Masschelein, C.A. *U.S. Patent: 4,957,749* **1990**.

95.   Blockmans, C.; Heilpron, M.; Masschelein, C.A. *Journal Am. Soc. Brew. Chem.* **1987**, *vol. 45*, pp. 85-90.

96.   Moll. M. *Journal Am. Soc. Brew. Chem.* **1990**, *vol. 48*, pp. 51-57.

97.   Takahashi, Y.; Takahashi, S.; Ujiie, F,; Sakuma, S.; Shimazu, T.; Kojima, K. *MBAA Technical Quarterly* **1991**, *vol. 28*, pp. 60-66.

98.   Bamforth, C.W.; Parsons, R. *Journal Am. Soc. Brew. Chem. 1985*, vol. *43, pp. 197-202.*

99.   Bamforth, C.W.; Boulton, C.A.; Clarkson, S.P.; Large, P.J. *Inst. of Brew., Austr. and N.Z. Section, Proc. of the 20th Conv.* **1988**, pp. 211-219.

100. Kronseder, H.; Schwarz, R.F. *MBAA Technical Quarterly* **1986**, *vol. 23*, pp. 1-5.

101. Peterson, A.; Evans, D. *Amer. Suppl. Inst. Food Symposium* **1991**, Dearborn, Michigan.

102. Vogelpohl, H. *Brauwelt* **1991**, *vol. 39*, pp. 1692-1704.

103. Arenda, C.W.; Lam. R. *U.S. Patent No: 4,847,696* **1989**.

104. Hooren van Heyningen, D.C.E.; van de Bergh, H.J.; van Strien, J.; Loggers, G.J. *Proceedings EBC Congress* **1987**, pp. 679-686.

105. Wisk, T.J.; Siebert, K.J. *Journal Am. Soc. Brew. Chem.* **1987**, *vol. 45*, pp. 14-18.

106. Teumac, F.N.; Ross, B.A.; Russoli, M.R. *MBAA Technical Quarterly* **1990**, *vol 27*, pp. 122-126.

107. Edens, L. *Eur. Pat. Applications 305,008 and 384,541* **1989 and 1990**.

108. Hofeldt, R.H.; White, S.A.C. *Eur. Pat. Applications 328,336 and 328,337* **1989**.

109. Teumac, F.N.; Zenner; B.D.; Ross, B.A.; Deardurff, L.A.; Rassouli, M.R. *Int. Patent: WO 91/17044* `**1991**.

RECEIVED May 10, 1993

Chapter 6

Monoterpenes and Monoterpene Glycosides in Wine Aromas

Seung K. Park and Ann C. Noble

Department of Viticulture and Enology, University of California, Davis, CA 95616

Varieties of grapes with intense floral aromas, such as Muscat of Alexandria, White Riesling, and Gewürztraminer, contain free monoterpenes along with glycosidically bound monoterpenes. Due to the hydrophilicity of bound monoterpenes they do not contribute to the wine aroma therefore winemakers are greatly interested in hydrolyzing these potential aroma precursors to release the free floral terpenes to enhance the varietal aroma. The increase of free and bound monoterpenes during development of grapes has been studied extensively, but little research has focused on the changes in free and bound monoterpenes during fermentation and subsequent aging. Recently, practical methods for hydrolysis of bound monoterpenes in wines have been explored. In this chapter, we will review techniques for analysis of free and bound monoterpenes, and discuss their changes during grape development, wine fermentation and aging. Application and problems associated with the use of glycosidase enzymes for hydrolysis of bound monoterpenes will also be discussed.

There is a growing interest in the study of glycosidically bound aroma compounds for their potential contribution to the varietal aroma of grapes, wines and other fruits. Of all wine volatiles, monoterpenes have been the most interesting compounds in "aromatic" grapes and wines because of their unique floral character. In 1946, Austerweil (1) first suggested that (free) terpenes contributed to the distinctive aroma of Muscat grapes. Changes in free monoterpene levels during maturation of grapes have been studied in Muscat (2), White Riesling and Cabernet Sauvignon (3), and Chenin blanc (4). In 1974, Cordonnier and Bayonove (5) speculated that bound glycosidic precursors of these terpenes occurred in grapes, when they found that terpenes increased during acidic or enzymatic hydrolysis of Muscat juice. Glycosidically bound monoterpenes from Muscat of Alexandria were first elucidated as a mixture of disaccharide glycosides of monoterpene alcohols by Williams et al. (6). Recognition of the importance of glycosidically bound monoterpenes as "flavor precursors" in grapes and wines has stimulated much research interest in the development of glycosidically bound terpenes during maturation to be able to harvest grapes at optimum maturity and quality.

The importance of monoterpenes in wine aroma

In studies of Muscat grape aroma, Ribéreau-Gayon et al. (7) suggested that linalool and geraniol were the most important compounds responsible for the varietal aroma of grapes. The role of monoterpenes in the aroma of Muscat grapes has been extensively studied over the past decade and the significance of these compounds clearly demonstrated (8). At present about 50 monoterpene compounds in *Vitis vinifera* grapes have been identified (8), of which the most abundant are geraniol, linalool, and nerol (Figure 1). Lower amounts of citronellol, nerol oxide, α-terpineol, diendiol-I, and various forms of linalool oxides have also been found. Seven major free terpenes were reported in Gewürztraminer juices and wines by Marais (9) including geraniol as the predominant terpene, followed by nerol, citronellol, diendiol-I, linalool, α-terpineol, and trace amounts of *trans*-furan linalool oxide. As shown in Table 1, linalool has a low sensory threshold compared to those of nerol and α-terpineol. Linalool oxides shown in Figure 1 are also present in most aromatic wines but at subthreshold levels and therefore they are unlikely to contribute to the aroma of grapes and wines.

Table 1. Sensory thresholds of major terpenes found in grapes and wines (μg/L)

	In sugar-water[a]	In wine[b]
Linalool	100	
Hotrienol		110
Geraniol	132	
Nerol	400	
a -Terpineol	460	
Linalool oxide (furan)		
trans -	>6,000	
cis -	>6,000	
Linalool oxide (pyran)		
trans -	3-5,000	
cis -	3-5,000	
Nerol oxide		110

[a] : Ribéreau-Gayon et al. (7).
[b] : Simpson (35).

Studies of flavor traditionally have focused on volatile compounds, but since Cordonnier and Bayonove (5) first suggested the presence of bound terpenes in grapes, more emphasis has been placed on analysis of nonvolatile precursors. The structures of conjugated terpenes were first identified by Williams et al.(6) as 6-*O*-β-L-rhamnopyranosyl-D-glucopyranoside (rutinoside) and 6-*O*-β-L-arabinofuranosyl-D-glucopyranoside. In addition to these disaccharides, Brillouet et al. (10) reported for the first time the presence of apiose [3-(hydroxymethyl)-D-erythrofuranose] in juice of Muscat Frontignan and Muscat of Alexandria grapes. The amount of apiose was almost equivalent (on the basis of relative proportions of terminal sugars) to the sum of rhamnosyl and arabinosyl glucosides. Previous studies have shown that these monoterpene glycosides accumulate in the berries to a greater extent than the free compounds (11, 12, 13), therefore, winemakers are very much interested in enhancing the floral aroma by hydrolyzing the bound monoterpenes. In addition to monoterpene alcohols, the polyhydroxylated monoterpenes have been studied with interest because of the significance of these compounds as potential flavorants (14, 15, 16). Several of the polyols, although odorless, are acid labile and readily form odor-producing compounds at ambient temperature and juice pH (14, 15, 17). Hotrienol, for example, appears to be formed wholly by acid catalyzed dehydration of diendiol-I (18, 19). By

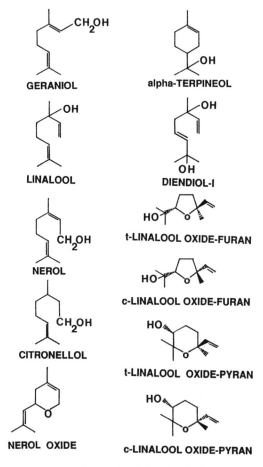

Figure 1. Major monoterpenes observed in Muscat of Alexandria grapes (Reprinted from ref. 64).

heating muscat juice at 70° C for 25 min, the concentrations of the furan linalool oxides, nerol oxide, hotrienol, and α-terpineol increased significantly. With the notable exceptions of α-terpineol, linalool, nerol, geraniol, and pyran linalool oxides, most of the heat-induced terpenes of the juices could be attributed to rearrangement products of grape polyols (14).

Analytical methods for free monoterpenes in grapes and wines

Continuous liquid/liquid extraction. The continuous liquid/liquid extraction method with Freon 11 (trichlorofluoromethane) has been used for enriching the volatile aroma compounds from grape berries and wines (13, 14, 20, 21) for several reasons. Freon 11 boils at a low temperature (23.7°C) so that formation of artifacts by heat can be minimized during extraction and subsequent concentration. Further, it recovers apolar compounds effectively while ethanol in wine is not extracted and other alcohols are weakly extracted. To determine aroma components in grapes, Rapp et al. (20), used 65% methanol to prevent enzymatic oxidation and hydrolysis during cell disruption. The homogenized sample was extracted with Freon 11 at 27°C for 15 hours. Up to 300 peaks were obtained using a 70 M glass capillary GC column. The reproducibility of this technique was studied by Marais (22) using a similar extractor. For 16 free terpenes, recoveries from grape juice and wine had coefficients of variation ranging from 1.2% to 13.9% and from 2.0% to 5.2%, respectively. Williams et al. (23) used Freon 11 for exhaustive extraction of polyhydroxylated monoterpenes (polyols) from Muscat of Alexandria grape juice after the free terpenes were removed with pentane. Reproducible quantitative results for the recovery of "terpene-diendiols" including diendiol-I, together with much smaller amounts of diols-II and III and hotrienol were obtained by adding pyridine to prevent adsorption to the sampling flask (18).

XAD-2 adsorbent. Gunata et al. (12) suggested a method of extraction and determination of both free and glycosidically bound terpenes and some aromatic alcohols using a non-ionic polymeric resin, Amberlite XAD-2. 84% to 93% of the free terpenes were recovered by elution with pentane. The bound fraction was then eluted with ethyl acetate, dried, redissolved and hydrolyzed with enzyme for GC analysis.

Dynamic Headspace Technique. Dynamic headspace techniques (eg. headspace analysis in which entrained volatiles are examined, versus static headspace analysis which simply samples a vapor equilibrated in a headspace of a closed container) have been widely used for the analysis of volatiles in numerous foods (24). There are many advantages in this technique over steam distillation (25) or porous polymer adsorption followed by heat-desorption. In addition to the simplicity of both apparatus and technique, several gas chromatographic analyses of the same headspace collection can be made. There are no problems with selectivity of porous resin adsorbents distorting the composition of collected headspace volatiles nor are artifacts from the decomposition of porous resins generated during thermal desorption of adsorbed volatiles. The method combines purified N_2 entrainment and continuous extraction with Freon 11 by slightly modifying the continuous liquid/liquid extractor first used by Rapp and Knipser (26). Over 99% of volatile compounds were recovered using this system with no artifacts or impurities produced under thess mild conditions. Reproducibilty, expressed as the coefficient of variation, ranged from 1% to 10% for a large number of compounds, while some aldehydes and some non-identified compounds gave values up to 30%.

Colorimetric method for total terpenes (rapid analysis). Dimitriadis and Williams (11) reported an analytical technique for the rapid analysis of total free volatile terpenes (FVT) and total potential volatile terpenes (PVT) which include terpene glycosides and polyols of grapes. The colorimetric method in which the amount of terpenes are expressed as "linalool equivalents" had originally been developed for determining the concentration of oxygenated terpenes, aldehydes, and esters in aqueous citrus essences (27). Steam distillation of juice at neutrality yielded free aroma compounds (FVT) of the grapes. This method can also be used for bound monoterpenes by acidifying the stripped juice at pH 2.0-2.2, and steam distilling the polyols and hydrolyzed terpenes to yield the PVT. Reaction of these distillates with a vanillin-sulfuric acid reagent produces a color, the intensity of which is proportional to the concentration of monoterpenes. Using different reference monoterpenes, it was demonstrated that the technique gave quantitative recovery of FVT from grape juice while recovery of the glycosides were less efficiently assayed (55% to 80%) than either free volatile monoterpenes or polyols. 2-phenyl ethanol, *n*-hexan-1-ol, and *cis*-hex-3-en-1-ol gave no positive reaction with the reagent, however, *trans*-hex-2-en-1-ol and allylic alcohols, which are present in grape juice and wine, interfere with the FVT determination. At 100 mg/L SO_2 had no effect on FVT figures, while very high concentrations of SO_2 (1000 mg/L) bleached the color, decreasing the apparent amounts of FVT and PVT.

Isolation methods for bound monoterpenes in grapes and wines

Because of the high molecular weight (MW 462 for linalyl rutinoside) and the high water solubility, the monoterpene glycosides of grapes have proven difficult to extract with solvents. Although Bitteur et al.(28) recently reported a method for direct analysis of terpene glycosides using reverse phase HPLC, most determinations of terpene glycosides are conducted indirectly by GC of the free terpenes liberated from the isolated glycosides. Previous methods such as gel-permeation and hydrophobic-interaction chromatography on polyacrylamide (Bio-Gel P-2) to purify (+)-neomenthyl-glucoside from peppermint leaf extracts (29) or solvent extraction followed by silica gel chromatography and preparative thin-layer electrophoresis (30) have proven to be inappropriate or unsuitable for the isolation of monoterpene glycosides from grape juice and wine because of the presence of large amounts of sugars in grapes (150-250 g/L) and in wines of alcohol (12%) or glycerol (often above 12g/L)(31, 32). Bound monoterpenes are analyzed primarily by three methods: C_{18} reverse-phase adsorbent (C_{18} RP), XAD-2 adsorbent, and distillation/colorimetric determination.

C_{18} Reverse-Phase Adsorbent. Trace enrichment of the organics from dilute aqueous solution on reverse-phase adsorbents was first proposed by Kirkland (33) and by Parliment (34) for flavor isolation. Williams et al. (31) developed the use of C_{18} RP liquid chromatography for the isolation of monoterpene glycosides and nor-isoprenoid glycosides from grape juice and wine. For juice analysis, a filtered sample is pumped down a glass column containing C_{18} RP. After being loaded, the column is washed with water to remove sugar and organic acids present in juice. Retained monoterpene glycosides are then eluted with methanol, and concentrated to dryness. The concentrate is washed with Freon 11 to ensure removal of any free terpenes prior to acid hydrolysis. The dried eluate is dissolved and free terpenes (liberated by enzyme hydrolysis) extracted with Freon 11 for GC analysis. A total of 56 peaks were identified including monoterpenes and C_{13} nor-isoprenoids. Using *p*-nitrophenyl-β-D-glucopyranoside (PNG), Park et al. (13) showed the recovery from C_{18} RP adsorbent to be 100%.

Amberlite XAD-2. Amberlite XAD-2 adsorbent has also been used for isolating the glycosidically bound terpenes in grapes and wine (12). With this procedure the bound fraction is directly eluted from the adsorbent using ethyl acetate after the free fraction has been eluted with pentane. This eluate is then dried and subjected to enzyme hydrolysis. The Amberlite XAD-2 resin displayed extraction capacities similar to those of C_{18} RP adsorbent. The recovery calculated using synthesized glycosides was between 90 and 100%. Schwab and Schreier (36) used the Amberlite XAD-2 technique to develop a simultaneous extraction and enzyme catalysis method to study flavor precursors of apple fruit. Glycosidic extracts were first isolated by Amberlite XAD-2 adsorption and eluted with methanol and dried. Enzymatic hydrolysis in buffer solution was conducted in a reaction tube for enzyme catalysis and subsequent liquid-liquid extraction. Because of simplicity and cheap price of Amberlite XAD-2 this method has been frequently used the isolation of bound aroma compounds in many other fruits and plants such as Lulo fruit (37), blackberry (38), red current leaves (39), fruits of *Prunes* species (40), pine apple (41), tomato (42), African mango (43).

Factors Influencing Concentration of Grape Monoterpenes

Many researchers have investigated the effect of grape maturity on terpene concentration in grapes and wines (2, 18, 44, 45, 46). Marais (9) suggested that lack of an intense characteristic aroma in many white cultivar wines in South Africa is caused in part to the high average temperatures during ripening of the grapes. In addition to the climatic effect on the formation of monoterpenes in grapes, there are many other viticultural and enological factors which also affect monoterpene levels and resulting wine aroma and quality. McCarthy (47) indicated that unirrigated Riesling vines produced fruit with higher potential volatile terpene concentrations (PVT) (eg. total concentration of glycosides and polyols) than that from irrigated vines. Reductions in crop load were also found to increase levels of PVT in irrigated treatments (47). Canopy manipulation in the form of vertical shoot training had little effect on PVT concentration despite an increase in fruit exposure. However, Williams et al. (48) in a preliminary study observed differences in terpene contents of Muscat Frontignac grapes cultivated with different trellis systems which afforded variation in light in the canopy. For several weeks after véraison, until the fruit reached a sugar level of about 15° Brix, all samples showed similar terpene content. Beyond this stage of maturity, the sun-exposed fruit had higher PVT levels. In addition to the effect of viticultural practices, there are a number of studies on the degradation of terpenoids in grapes by *Botrytis cinerea* (17, 49, 50, 51, 52, 53). Boidron (49) showed that there was a decrease in the concentration of acyclic monoterpene alcohols and oxides in grape juice which was fortified with these compounds and then inoculated with *B. cinerea*. After 25 days of fungal growth, none of the added monoterpenes could be detected in the juice. In addition a number of products are formed by *B. cinerea* from terpenes such as linalool (54) and citronellol (55). Shimizu (56) reported that *B. cinerea* did not produce terpenes in grapes without starting terpene products, but actively transformed linalool into other monoterpenes. Perhaps the major change in terpene concentration occurs as a function of berry development. Generally, an increase in grape maturity yields an increase in the concentration of free and bound terpenes, but differences are observed between varieties and for different compounds. *Trans*-furan linalool oxide, α–terpineol, and citronellol increased in Gewürztraminer juice as the grapes ripened, while citronellol and *trans*-geranic acid increased in wines made from grapes of increasing ripeness. However, linalool and diendiol-I decreased significantly with grape maturation (9). Several studies also showed that an increase in linalool occurs with grape maturation, followed by a decrease at some ripening stage, usually at over-ripeness (46, 57). In a study of free terpenes in Muscat of Alexandria grapes, linalool was first detected two weeks after the onset of sugar accumulation (véraison) and increased steadily as long as the grapes were left on the vine (2). Other volatiles of possible significance in the aroma of the ripe grapes appeared 2-4 weeks after linalool was first observed and then increased sharply. Similar data were reported by Wilson et al. (18) in developing Muscat grapes. However, free nerol and geraniol, rapidly decreased about two weeks before véraison then almost disappeared toward harvest, while the bound forms of nerol and geraniol steadily increased with sugar accumulation. Contrary to this, Gunata et al. (46) reported free and bound nerol and geraniol steadily increased even after ripeness was reached, whereas free linalool decreased slowly after ripeness while bound linalool decreased rapidly. In a developmental study of Muscat of Alexandria, major free and bound monoterpenes (linalool, geraniol, and nerol) steadily increased over the growing season and continued to rise after the maturity level at which commercial harvest occurred (13). Consistent with Marais's (9) speculation that high temperatures interfered with terpene accumulation, levels of free and bound terpenes dropped in the Muscat grapes following several very hot days >100°F (13). Difference in results between France, California and Australia is most likely due differences in maturity at harvest and in environmental factors, such as soil and temperature.

Distribution of free and bound monoterpenes in grapes

Understanding the sites of synthesis and storage of free and bound monoterpenes in grape berries is of primary importance for practical purposes. For example, in grapes used for winemaking, knowledge of the concentration as well as the distribution of free and bound monoterpenes in different sites in the berries would offer a valuable guide in selecting skin contact and press conditions to optimize the aroma concentration in juice. Bayonove et al. (58) observed a highly uneven distribution of some free monoterpenes in different fractions of Muscat grapes. Geraniol and nerol, for example, were associated primarily with the skins of the berries, whereas linalool was more uniformly distributed between the juice and the solid parts of the fruit. Similar results were also obtained by Cordonnier et al. (59, 60): linalool was almost equally distributed between juice and skin, whereas 95% of geraniol and nerol were found in the skin of Muscat of Alexandria grapes. Since the aromatic profile of the skin differs considerably from that of juice, the intensity and quality of the aroma of wine might be influenced by different skin contact times (61, 62). Gunata et al. (46) studied the distribution of free and bound fractions in the skins, pulp, and juices of Muscat of Alexandria and Frontignan. Generally, bound forms of terpenes were more abundant, except for free linalool which was more abundant in the skins of Muscat of Alexandria. A total of 3571 µg/kg bound terpenes and 2904 µg/kg free terpenes were found in the skins, respectively accounting for 57% and 85% of the total bound and free terpenes in the whole berries. From their study, linalool, geraniol, and nerol were the most abundant compounds in both varieties. Wilson et al. (63) also studied the distribution of free and bound terpenes in Muscat of Alexandria, Frontignac, and Traminer. In all three varieties, most of the free geraniol and nerol were associated with the skins of the berries and free linalool was evenly distributed between skins and juice of the two Muscat varieties (Frontignac and Alexandria). In Traminer, free and bound linalool and diendiol-I were found in low concentrations (less than 20 µg/kg and 50 µg/kg for free and bound linalool and diendiol-I, respectively). However, levels of free and bound linalool and diendiol-I were higher in Frontignac and Alexandria than Traminer and were uniformly distributed between skin and juice. For Muscat of Alexandria grapes in a warm region of California (13), about 90% of monoterpenes were glycosidically bound, while only 10% were free. The distribution of free and bound monoterpenes between the skins and mesocarp (pulp and juice) changed constantly during ripening of the berries. At harvest, 4.6% and 5.9% of the three major monoterpenes (linalool, geraniol, and nerol) occurred as free monoterpenes in the skin and mesocarp, respectively, whereas 31% and 59% of total monoterpenes were found as glycosides in the skin and mesocarp. Considering the weight percentages of the skins in whole berry (10 to 15%) the high proportion (31%) of bound monoterpenes in the skins suggests the possibility of using greater press force and extended skin contact during winemaking to increase the extraction of bound monoterpenes from grapes.

Changes of Monoterpene During Fermentation and Aging

The aroma of wine is a function of the compounds which are extracted unmodified from the grapes and those formed during fermentation by yeast. Further changes in volatile composition, including in free and bound monoterpenes, occur during processing and aging steps. Total free monoterpenes in Muscat of Alexandria wine have been shown to increase, with a corresponding reduction in the total bound monoterpenes during fermentation and subsequent storage due to hydrolysis in the acidic wine system (64, 65). Of the major free monoterpenes in Muscat of Alexandria, geraniol present in must at 49 μg/L decreased to 5 μg/L after fermentation then reappeared slowly as the aging proceeded. This slow reappearance is mainly due to the hydrolysis of bound geraniol in wine. Free α-terpineol, initially present as a minor terpene in must (2 μg/L) increased slowly during fermentation to 5 μg/L and further increased to 70 μg/L in wine after aging for 13.5 months at 10 °C. The increase during storage of α-terpineol above the amount present in must as the glycoside (31 μg/L) is the result of the acid-catalyzed isomerization of other free monoterpenes such as linalool, geraniol, and nerol (66). In our on-going research, when Gewürztraminer wine was stored at 15 °C for 3 months, 27% of bound terpenes were hydrolyzed (74). As shown in Figure 2, the three major glycosides in Muscat of Alexandria wine are hydrolyzed at different rates, with linalool hydrolyzed most rapidly, followed by geraniol and nerol. For this wine, storage at 10°C will result in complete hydrolysis of the linalool glycoside within 1.7 years whereas it would take 4.5 years to deplete the geraniol glycoside in wine (65).

Enhancing Wine Aroma: Hydrolysis of Glycosidically Bound Monoterpenes

Acidic hydrolysis. Acid hydrolysis of grape glycosides has been studied as a method for releasing the bound monoterpenes, however, acid hydrolysis, promoted by heating, results in rearrangement of the monoterpene aglycones (14, 23). Thermal induction of volatile monoterpenes in juices of Muscat grapes was studied by GC and GC-MS of headspace samples (14) and the roles of the four grape polyols upon heating were also investigated. Heating juice for 15 min at 70°C significantly increased the concentration of the furan linalool oxides, nerol oxide, hotrienol, and α-terpineol. In addition to these compounds, other terpenes and terpene derivatives, such as 2,6,6-trimethyl-2-vinyltetrahydropyran, myrcen-2-ol, and other terpene-derived compounds, all previously unrecognized as grape products, were found (14). Therefore, pH adjustment of juices, heat-treatment for pasteurization purposes, or even extensive storage periods will induce changes in the concentration of volatile terpene compounds. Hydrolysis at pH 1 of precursor fractions from grapes yielded a very different pattern of volatiles than that seen at pH 3 (31). However, it has been observed that prolonged heating even at pH 3 ultimately altered the sensory character by imparting a eucalyptus-like aroma, attributable to the presence of an excess quantity of 1,8-cineole in the headspace composition of the juice (67). Although acidic hydrolysis is not practical nor desirable for enhancing aroma in premium wines, brief heating of a portion of wine at 85°C for 2 minutes followed by immediate cooling by heat exchanger has been used in production of inexpensive wines in Australia.

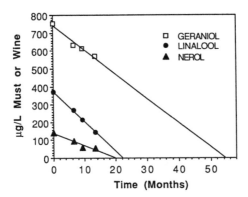

Figure 2. Change in concentration of terpene glycosides during storage of Muscat of Alexandria wine. Initial grape must concentration shown at time 0 (65).

Enzymatic hydrolysis. Addition of an enzyme hydrolyzate of the bound fraction of Muscat wine produced a significant difference in wine aroma, illustrating the potential effectiveness in enhancing wine aroma (68). Various enzyme preparations have been used to hydrolyze the bound fraction in grapes (6, 12, 69), but all of these studies involved isolation of monoterpene glycosides before enzyme hydrolysis because of inhibition of enzyme activity by glucose and ethanol, present in grapes or in wine. Bayonove et al. (70) demonstrated the presence of a weak natural glycosidase activity in Muscat grape juice, which resulted in an increase in the concentration of nerol and geraniol in juice adjusted to pH 5 and incubated at 30°C for 24 hr. Shorter holding times at lower temperatures or lower pH values gave smaller increases while no increase in free terpenes was produced in pasteurized juice. The enzyme, which was located primarily in the juice fraction, increased in activity with berry maturity. Glucosidase from grapes or other sources is strongly inhibited by the high sugar concentration of grape musts (69, 71, 72). For example, at 50mM of glucose, 50% inhibition of the activity of grape glucosidase occurred (69), hence neither grape nor commercial enzymes, such as Rohapect C (Röhm Gmbh, Germany) or other glucosidic enzymes can be used directly in juice. However, Rohapect C, a crude pectinase with glucosidase activity, is relatively insensitive to ethanol in wine (69), while activity of an endo-glucosidase isolated from *Aspergillus niger* is enhanced in the presence of ethanol up to a maximum at 9% (72) suggesting the possibility of use in wine. Addition of a commercially available glucanase with glucosidase activity to Morio-Muscat wine directly after fermentation was shown to double the concentration of free geraniol and nerol (73). Our unpublished results on the hydrolysis of terpene glycosides have shown that some commercially available enzyme preparations, (e. g., Rohapect 7104 and Novoferm 12L) hydrolyze the terpene glycosides in finished wines completely. The rate of hydrolysis depends on the amount of enzyme added to the wines, type of terpene glycosides, acidity of wine, amount of residual sugar and ethanol present in finished wines, and storage conditions upon enzyme addition. Although enzymatic hydrolysis seems to be very promising to enhance the varietal aroma in wines, there are several problems associated with enzyme applications: first, since most effective commercial enzymes either crude pectinase or pectinase containing β-glycosidase are prepared from fungal source (*Aspergillus niger*) so they may change the color of wines as a result of polyphenol oxidase (PPO) activity in the presence of oxygen. Second, the enzyme added to the finished wine can be precipitated during storage of wine.

Summary

New grape growing and wine making technology has been developed during the last decade which have been used to improve the quality of wines. Identification of monoterpene glycosides as potential flavor compounds has stimulated research on factors influencing their accumulation during fruit ripening, on their localization and on potential enhancement of wine aroma by their hydrolysis. Although the traditional methods of winemaking are still the most reliable for production of fine wines, considerable research is being directed toward utilizing these glycosides to improve wine flavor.

Literature Cited

1. Austerweil, G. *Ind. Perfume.* **1946**, 1,195.
2. Hardy, P. J. *Phytochem.* **1970**, 9,709.
3. Egorov, I. A.; Rodopulo, A. K.; Bezzubov. A. A.; Skiripnik,Y. A.; Nechaev. L. N. *Prikl. Biokhvn. Mikrobiol.* **1978**, 14,109.
4. Augustyn, O. P. H.; Rapp, A. *S. Afr. J. Enol. Vitic.* **1982**, 3, 47.
5. Cordonnier, R.; Bayonove, C. *C. R. Acad. Sci.* **1974**, 278,3387.
6. Williams, P. J.; Strauss, C. R.; Wilson, B.; Massy-Westropp, R. A. *Phytochem.* **1982**, 21, 2013.
7. Ribereau-Gayon, P.; Boidron, J. N.; Terrier, A. *J. Agric. Food Chem.* **1975**, 23, 1042.
8. Strauss, C. R.; Wilson, B.; Gooley, P. R.; Williams, P. J. *In Biogeneration of Aromas*; Parliment, T. H.; Croteau, R. eds.; ACS Symposium series no. 317; American Chemical Society: Washington DC, **1986**; pp 222-242.
9. Marais, J. *Vitis.* **1987**, 26, 231.
10. Brillouet, J.-M.; Gunata, Z.; Bitteur, S.; Cordonnier, R. *J. Agric. Food Chem.* **1989**, 37, 910.
11. Dimitriadis, E.; Williams, P. J. *Am. J. Enol. Vitic.* **1984**, 35,66.
12. Gunata, Z.; Bayonove, C. L.; Baume, R. L.; Cordonnier, R. E. *J. Chromatogr.***1985**, 331, 83.
13. Park, S. K.; Morrison, J. C.; Adams, D. O.; Noble, A. C. *J. Agric. Food Chem.* **1991**, 39, 514.
14. Williams, P. J.; Strauss, C. R.; Wilson, B. *J. Agric. Food Chem.* **1980**, 28, 766.
15. Williams, P. J.; Strauss, C. R.; Wilson, B. *Phytochem.* **1980**, 19, 1137.
16. Rapp, A.; Mandery, H.; Ullemeyer, H. *Vitis.* **1983**, 22, 225.
17. Rapp, A.; Mandery, H.; Niebergall, H.*Vitis.* **1986**, 25, 79.
18. Wilson, B.; Strauss, C. R.; Williams, P. J. *J. Agric. Food Chem.* **1984**, 32, 919.
19. Engel, K.-H.; Tressl, R. *J. Agric. Food Chem.* **1983**, 31, 998.
20. Rapp, A.; Hastrich, H.; Engel, L. *Vitis.* **1976**, 15, 29.
21. Etievant, P. X.; Bayonove, C. L. *J. Sci. Food Agric.* **1983**, 34, 393.
22. Marais, J. *S. Afr. J. Enol. Vitic.* **1986**, 7,21.
23. Williams, P. J.; Strauss, C. R.; Wilson, B. *Am. J. Enol. Vitic.* **1981**, 32, 230.
24. *Analysis of foods and beverages, headspace techniques*; Charalambous, G. Ed.; Academic Press: New York, **1978**, pp 1-390.
25. Schmaus, G.; Kubeczka, K.-H. *In Essential oils and aromatic plants;* Svendson, B, A.; Scheffer, J. J. C., Ed.; Dr. W. Junk Publisher: Dordrecht, The Netherlands, **1985**, p. 127.
26. Rapp, A.; Knipser, W. *Chromatographia.* **1980**, 13, 698.
27. Attaway, J. A.; Wolford, R. W.; Dougherty, M. H.; Edwards, G. J. *J. Agric. Food Chem.* **1967**, 15, 688.

28. Bitteur, S.; Gunata, Z.; Brillouet, J.-M.; Bayonove, C. *J. Sci. Food Chem.* **1989,** 47, 341.
29. Croteau, R.; El-Hindawi, S.; El-Bialy, H. *Anal. Biochem.* **1984,** 137, 389.
30. Francis, M. J. O.; Allock, C. *Phytochem.* **1969,** 8, 1339.
31. Williams, P. J.; Strauss, C. R.; Wilson, B.; Massey-Westropp, R. A. *J. Chromatogr.* **1982,** 235, 471.
32. Peynaud, E. *In Knowing and making wine*; John Wiley & Sons: New York, **1984**; pp 36-37.
33. Kirkland, J. *Analyst* (London). **1974,** 99, 859.
34. Parliment, T. H.; *J. Agric. Food Chem.* **1981,** 29, 836.
35. Simpson, R. F. *In Food Technol. in Australia.* **1979,** 516.
36. Schwab, W.; Schreier, P. *J. Agric. Food Chem.* **1988,** 36, 1238.
37. Suarez, M.; Dugue, C.; Wintoch, H. Schreier, P. *J. Agric. Food Chem.* **1991,** 39, 1643.
38. Humpt, H.-U.; Schreier, P. *J. Agric. Food Chem.* **1991,** 39, 1830.
39. Humpt, H.-U.; Winterhalter, P.; Schreier, P. *J. Agric. Food Chem.* **1991,** 39, 1833.
40. Winterhalter, P.; Schwab, M.; Schreier, P. *J. Agric. Food Chem.* **1991,** 39, 778.
41. Wu, P.; Kuo, M-C.; Hartman, T. G.; Rosen, R. T.; Ho. C-T. *J. Agric. Food Chem.* **1991,** 39, 170.
42. Marlatt, C.; Ho, C-T.; Chien, M. *J. Agric. Food Chem.* **1992,** 40, 249.
43. Adedeji, J.; Hartman, T. G.; Lech, J.; Ho, C-T. *J. Agric. Food Chem.* **1992,** 40, 659.
44. Bayonove, C.; Cordonnier, R. *Ann. Technol. Agric.* **1970,** 19, 79.
45. Rapp, A.; Hastrich, H.; Engel, L.; Knipser, W. *In Flavor of food and beverages*; Charalambous, G.; Ingeltt, G. E. Ed.; Academic Press: New York, **1978**; pp 391-417.
46. Gunata, Y. Z.; Bayonove, C. L.; Baumes, R. L.; Cordonnier, R. E. *J. Sci. Food. Agric.* **1985,** 36, 857.
47. McCarthy, M. G.; *In Influence of irrigation, crop thinning, and canopy manipulation on composition and aroma of Riesling grapes.* Ph. D. Thesis. **1986.** Waite Agricultural Institute, University of Adelaide. Australia.
48. Williams, P. J.; Strauss, C. R.; Wilson, B.; Dimitriadis, E. *In Topics in flavor research*; Berger, R. G., Nitz, S., Schreier, P. Ed.; H. Eichhorn: Marzling-Hangenham, W. Germany, **1985**; pp 335-352.
49. Boidron, J. N. *Ann. Technol. Agric.* **1978,** 27, 141.
50. Shimizu, J.; Uehara, M.; Watanabe, M. *Agric. Biol. Chem.* **1982,** 27, 141.
51. Bock, G.; Benda, I.; Schreier, P. *Appl. Microbial. Biotechnol.* **1988,** 27, 351.
52. Brunerie, p.; Benda, I.; Bock, G.; Schreier, P. *In Bioflavor'87*; Schreier. P. Ed.; Walteer de Gruyter: Berlin, **1988,**; pp 435-443.
53. Rapp, A.; Mandery, H. *In Bioflavor'87*; Schreier. P. Ed.; Walteer de Gruyter: Berlin, **1988,**; pp 445-452.
54. Bock, G.; Benda, I.; Schreier, P. *J. Food Sci.* **1986,** 51, 659-662.
55. Brunerie, P.; Benda, I.; Bock, G.; Schreier, P. *Appl. Microbial. Biotechnol.* **1987,** 27, 6.
56. Shimizu, J. *Agric. Biol. Chem.* **1982,** 46, 2265.
57. Marais, J.; van Wyk, C. J. *S. Afr. J. Enol. Vitic.* **1986,** 7, 26.
58. Bayonove, C.; Cordonnier, R.; Ratier, R. *C. R. Acad. Agric. Fr.* **1974,** 60, 1321.
59. Cordonnier, R.; Bayonove, C. *Parfum. Cosmet. Aromes.* **1978,** 24, 67.
60. Cordonnier, R.; Bayonove, C. *Vigne. Vin.* **1981,** 15, 269.
61. Lamikanra, O.; Garlick, D. Food Chem. 1987, 26, 245.
62. Marais, J.; Rapp, A. *S. Afr. J. Enol. Vitic.* **1988,** 9, 22.
63. Wilson, B.; Strauss, C. R.; Williams, P. J. *Am. J. Enol. Vitic.* **1986,** 37, 107.

64. Park, S. K. Distribution of free and bound forms of monoterpenes during maturation of Muscat of Alexandria and Symphony grapes. *MS Thesis*, University of California, Davis. **1989**.
65. Park, S. K.; Morrison, J. C.; Admas, D. O. University of California, Davis, *unpublished data.*
66. Rapp, A. H.; Mandery, H.; Guntert, M. *In Proceedings of the alko symposium on flavor research of alcoholic beverages*; Nykanen, L., Lehtonen, P., Ed.; Foundation for Biotechnical and Industrial Fermentation Research: Helsinki, Finland, **1984**; pp 225-274.
67. Williams, P. J.; Strauss, C. R.; Wilson, B.; Massy-Westropp. *J. Agric. Food Chem.* **1982**, 30, 1219.
68. Noble, A. C.; Strauss, C. R.; Williams, P. J.; Wilson, B. *In Flavor Science and Technology*; Marten, M., Dalen, G. A., Russwurm Jr., H., Ed.; John Wiley & Sons, New York, **1987**; pp 383-390.
69. Aryan, A. P.; Wilson, B; Strauss, C. R.; Williams, P.J. *Am. J. Enol. Vitic.* **1987**, 38, 182.
70. Bayonove, C.; Gunata, Z.; Cordonnier, R. *Bull. Olv.* **1984**, 643, 741.
71, Großmann, M.; Rapp, A.; Rieth, W. *Deutsche Lebensmittel-Rundschau.* **1988**, 84, 35.
72. Shoseyov, O.; Bravdo, B.-A.; Ikan, R.; Chen, I. *Phytochem.* **1988**, 27, 1973.
73. Großmann, M.; Rapp, A.; *Deutsche Lebensmittel-Rundschau.* **1988**, 84, 35.
74. Robichaud, J.; Park, S. K. Enzyme modification of aroma of Gewürztraminer wine. *Presented at the American Society for Enology and Viticulture annual meeting*, Reno, Nevada, June 25-27, **1992**, U. S. A.

RECEIVED May 5, 1993

Chapter 7

Brettanomyces and *Dekkera*
Implications in Wine Making

K. C. Fugelsang, M. M. Osborn, and C. J. Muller

Department of Enology, Food Science, and Nutrition and Viticulture and
Enology Research Center, California State University,
Fresno, CA 93740–0089

Yeasts of the genera *Brettanomyces* and *Dekkera*
pose a serious threat to premium wine production.
It has been conservatively estimated that annual
economic losses resulting from their growth run
into the hundreds of thousands of dollars. The
historically ascribed habitat for both yeasts is
barrel-aging red wine. Research presented here
points to the potential for significant activity
during fermentation. Fermentative phase growth
not only results in formation of objectionable
metabolites that develop and intensify during
aging, but also negatively impacts activity of the
wine yeast *Saccharomyces*. Efforts should be
directed toward exclusion of the organism and in
the case of already established infections,
monitoring, isolation and control. Despite valid
concerns regarding "Brett"-growth in wine, some
creative winemakers are currently exploring and
advocating selective and controlled utilization of
these yeasts as stylistic tools.

Background

The process of winemaking, from vineyard to bottled
product, reflects not only the unique contribution of the
grapes and winemaker, but the combined activities of
resident vineyard and winery flora as well. Indeed, if the
process were not controlled, we would see a succession of
microbial populations (including both yeast and bacteria)
representing, initially, those species present on the fruit
and later those species tolerant of higher alcohol
environments.

Historically, winemaking has relied upon mixed culture
fermentations resulting from a succession of yeast

0097–6156/93/0536–0110$06.00/0

populations potentially representing several genera. Such
fermentations are initiated by weakly fermentative species
which are relatively alcohol sensitive.   These are quickly
overgrown   by   strongly   fermentative   strains   of
Saccharomyces.   The details regarding identification and
enumeration of native yeasts can be found in   various
reports (*15, 21, 23*).

Distribution and population densities of native
species   vary   widely,   reflecting   environmental   and
viticultural factors as well as maturity and integrity of
the fruit.   Reed and Nagodawithana (*23*), citing French
research, report yeast cell counts ranging from less than
160 to 10^5 CFU/berry.   Worldwide, the most frequently
isolated native yeasts are *Hanseniaspora uvarum* and its
asexual or "imperfect" counterpart *Kloeckera apiculata*.
Other common native yeasts are *Metschnikowia pulcherrima*
and its imperfect form, *Candida pulcherrima* (*23*).   Yeasts
isolated less frequently include *Torulopsis delbrueckeii*,
Hansenula   anomala,   and   *Pichia   membranefaciens*.
Saccharomyces sp. are infrequently isolated from vineyards
where   winery   wastes   are   not   reincorporated   as   soil
amendments.   In   instances   where   such   practices   are
utilized, resident populations may be high.   Pardo *et.
al.*, (*17*) report levels as high as 5 x 10^5 CFU in Spanish
grape musts.   These same workers also report incoming
musts where *Saccharomyces* was not isolated, suggesting
considerable habitat variability.

In California, there is renewed interest among some
winemakers in native yeast fermentation. However, it is our
opinion   that,   aside   from   isolated   locales   where
viticultural practices artificially create a microbial
community with relatively high population density of
Saccharomyces, native vineyard yeasts represent species
that are, with few exceptions, ill suited for winemaking.
At best, most of the "wild" yeast flora is represented by
weak fermenters, capable of producing only 4-5% alcohol,
concomitant with a variety of unpleasant metabolites.
Fortunately for winemakers (past and present) working under
these conditions, one or more strains of more strongly
fermentative and alcohol tolerant *Saccharomyces* eventually
predominates and completes fermentation (*42*).

Early vintners recognized the importance of yeast in
fermentation and made conscious efforts to propagate those
strains they believed important in the process. The extent
of their efforts no doubt varied with scientific awareness,
ranging   from   plowing   fermented   pomace   back   into   the
vineyard to maintenance of pure culture strains in the
winery laboratory.   Needless to say, product uniformity
represented a serious problem to early winemakers and most
of the world's wine was likely produced more by fortuitous
circumstance than intentional direction.   With a better
understanding of the elements involved in fermentation as
well as the technology for implementation and manipulation,
winemakers in the last century (and particularly the last
50 years) have gained a greater control over the onset and

outcome of fermentation as well as cellaring and bottling wine.

To produce a consistently superior product on a global basis, winemakers needed yeasts with relatively uniform and predictable microbiological properties. Chief among these was sustainable viability over the course of fermentation. This required the ability to grow in the low pH (<3.8) and (initially) high osmotic pressure environment of grape juice and tolerance of increasing levels (12-14% vol/vol) of alcohol. These requirements, by themselves, eliminate most of the native yeast flora present in the vineyard. Further attributes of wine yeasts include rapid and complete conversion of sugar to alcohol with minimal production of other by-products, as well as capabilities of fermenting at low (<60°F) temperatures and relative resistance to the commonly used antioxidant and preservative, sulfur dioxide. Since rapid clarification was essential to processing for the increasing numbers of large wineries, agglomeration properties were also sought. Among the various yeasts that grow on fruit and during the process of winemaking, only species and strains of *Saccharomyces* combined all the necessary requirements.

One of the more important technical advances in the last 30 years was development and routine production and marketing of dehydrated wine yeast. Since the first release of a single strain by Red Star Yeasts (Universal Foods) in 1965, the selection of routinely available, reasonably-priced dehydrated yeasts has grown to include over a dozen strains marketed internationally by several companies.

Use of wine active dry yeasts (WADY), as the product came to be called, gained rapid and wide acceptance in the U.S. In winegrowing areas with a much longer history of winemaking (hence already established populations of wine yeast), acceptance was much slower. Owing in large part to production of new specialty strains, today virtually all winemaking areas of the world utilize active dry yeasts.

Currently, concerns regarding control of several groups of wine microorganisms have re-emerged as important issues among winemakers. This in part stems from the wine industry's increased awareness of and interest in reducing the levels of sulfur dioxide used in processing. Aside from perceived advantages in certain winemaking applications, increasing concern from public and medical fields has led to enactment of disclosure requirements for bottled wine. The U.S. Government now requires label disclosure on wines containing more than 10 mg/L total sulfites. Facing the probability of further restrictions and, potentially, elimination of the compound all together, winemakers are using much lower levels of sulfur dioxide during processing. This practice, no doubt, has contributed to the proliferation and spread of spoilage microbes.

In the case of wine bacteria, predictable control and utilization of the malolactic fermentation has been and still is a major area of interest and concern to the winemaking community (9). Recent reviews are available (3, 4). Likewise, growth of acetic acid bacteria still represents a problem to the world wine producers. Excellent reviews of this subject have also been published recently (5, 6).

Two of the more troublesome yeasts that grow in juice and wine are *Brettanomyces* sp. and its sporulating equivalent *Dekkera* sp. Historically, the wine community has viewed *Brettanomyces* as producing frank spoilage in wines where it could grow. Conservative loss estimates range into the hundreds of thousands of dollars annually, not only from overtly spoiled unmarketable wines, but also wines of diminished quality that do not command their expected market price.

Among wine professionals working with *Brettanomyces* and *Dekkera*, it was observed that despite the negative connotations surrounding these yeasts, some (not all) internationally recognized, award winning wines had perceivable "Brett character." The question that arose from these observations was then, is some "Brett character" beneficial in certain styles of wine?

The answer to this question is clouded by several problems dealing with routine recovery and identification of the yeast. Much of what is known or believed about "Brett," from the winemaker's point of view, is based upon sensory changes occurring in the wine. In surprisingly few instances are these observations supported by laboratory validation or conclusive identification of the organism(s) involved. In the above example of supposed "Brett-taint" in award-winning wines, subsequent attempts to recover and culture the suspect microbe proved unsuccessful.

Various sensory descriptors have been used to characterize *Brettanomyces*/*Dekkera*-tainted fermentations and their resultant wines. These include "cider" and "clove-like," or "spicey," "smokey," "medicinal" and "mousey." Other frequent descriptors include "horsey," "wet wool" and even "burnt beans." Further, the sensory effects of "Brett-like" yeast growing in wine appear to be different in different wine types. Not only are significant sensory differences seen when reds and whites are compared, but within either group, sensory interpretation varies widely. In one instance, "Brett" contamination may result in odors variously described as "spicey" or "medicinal," whereas in others, activity of the yeasts produces odors reminiscent of "rodent-cage litter."

From what has been said, it is apparent that care must be taken when applying any of these descriptors solely, and without laboratory validation, to implicate "Brett" or *Dekkera* growth in wine. For example, "mousiness" is commonly used to describe *Brettanomyces*/*Dekkera*-tainted wines. However, this may also describe wines where certain strains of offensive heterofermentative lactobacilli have grown. In both cases, the compound eliciting the unpleasant odor is the same, a substituted

tetrahydropyridine derived from the amino acid lysine and
ethanol (2, 11, 12). Thus, attempts to gather information
on the sensory properties of suspect yeast (without
suitable prior identification) are difficult to interpret.
There is more than one recognized species which may
occur in wine or juice. Van der Walt (30, 31) describes
seven species of Brettanomyces and two species of Dekkera.
Subsequently, Brettanomyces was expanded to include 9
species while Dekkera remained unchanged (32). Of those
species described, only B. intermedius and B. lambicus were
originally isolated from grape wines or juice. In
characterizing 57 isolates, Smith (personal communication,
1992) reports identification of only B. custersii and D.
intermedia.

Another question arises from sensory examination in the
absence of definitive laboratory identification. Are
these differences the result of the same organism growing
in different wine varieties, or are we dealing with two or
more different yeasts?

Part of the problem of laboratory correlation of
sensory impressions lies in the difficulties encountered in
routine isolation and identification at the genus level.
Most protocols for isolation of Brettanomyces or Dekkera
from natural sources (fermenting juice or clarifying wine)
use selective inhibitors such as actidione (cycloheximide)
to impede the growth of numerically superior (but
inhibitor-sensitive) microbes such as Saccharomyces sp.

Use of actidione suffers potentially from two
interpretational problems. First, in preparation of growth
media utilizing the agent, laboratory personnel typically
incorporate 20-50 mg/L into agar/broth before autoclaving
in the expectation that they have sufficient quantities of
the active form after the sterilization cycle is complete.
Depending upon the age and condition of the actidione, this
may or may not be a valid assumption. Secondly,
Brettanomyces and Dekkera are not the only yeasts present
on fruit and in fermentation that are resistant to the
effects of the inhibitor. The property is also common to
Hanseniaspora, Kloeckera, and Schizosaccharomyces (30), all
of which are part of the normal flora of grapes and early
stages of wine fermentations. Thus, the fact that yeast
colonies develop after 7-10 days of incubation on
laboratory media should not be taken as confirmation for
"Brett" or Dekkera. Follow up screening is recommended.
Certainly in the case of those facilities planning serious
investigation, it is necessary to separate Brett. from
Dekkera.

Species of both Brettanomyces and Dekkera are strongly
acidogenic, producing large amounts of acetic acid from
growth on glucose (see Table 1). Formation of acetic acid
is believed to result from oxidation of ethanol rather than
via pyruvate (24). Both Brett. and Dekkera may produce
amounts of acetic acid sufficient to inhibit and eventually
kill cultures maintained on unbuffered substrate. Thus,
routine laboratory maintenance media contains 2% (w/v)

Table 1 Comparison of Fatty Acid Levels
in Pure- and Co-Culture Fermentations

Yeast(s)	Acetic Acid (mg/L)	Octanoic Acid (mg/L)	Decanoic Acid (mg/L)
Saccharomyces	82	18	7
Sacc. x Brett.	249	92	16
Sacc. x Dekk.	672	93	11

calcium carbonate.   Accumulation of acetic acid is reported to result from low level activity of TCA Cycle enzymes as well as an imbalance in reduced/oxidized states of the coenzyme involved in oxidation of ethanol to acetic acid (24).   Additional fatty acid metabolites that potentially contribute to the sensory profile of tainted wine include isobutyric, isolvaleric and 2-methyl-butyric acids (36, 37).

Formation of acetic acid is not solely diagnostic for either yeast. *Hansenula anomala* which is also present as part of the native vineyard flora (and thus may be found in the early stages of fermentation) also shares this property.   Upon isolation, most labs rely on microscopic comparison of cell morphology as part of their identification. While cell shape certainly plays a role in identification, frequently too much emphasis is placed upon this criterion. Yeasts in general exhibit variable cell morphology depending upon age, culture medium and environmental stress. For example, *Brettanomyces* grown on solid agarose substrate may appear considerably different from *Brettanomyces* isolated in barrel aging wine. Classically both genera and their respective species exhibit cell shapes described as "ogival." Reminiscent of Gothic arches, ogival cell morphology results from the restricted polar budding characteristic of the yeast. Hence it would be expected of older cells in the population. Thus, while a useful characteristic, relatively few cells (generally less than 10%, depending upon age and environmental conditions of culture) in the population may exhibit this shape (27). Van der Walt (30) also cautions that while ogival shape is characteristic of *Brettanomyces* or *Dekkera*, it is not exclusive to the genera.

Probably the most significant stumbling block in successful routine laboratory identification of *Brettanomyces* and *Dekkera* lies in the fundamental requirements of taxonomic guides to demonstrate the presence (or absence) of a sexual phase in the life cycle of the yeast. Mycologists utilize a system of classification, much like any other used in biology, which is based upon the degree of similarity between organisms. In the case of fungi (in this case yeasts) however, classification (to the genus level) requires demonstration of the presence or absence of a sexual phase in the life cycle. Yeasts regularly reproduce by the asexual process called budding. This mode of reproduction, which may occur repeatedly for many generations, is familiar to all who have observed yeasts during fermentation or growing on laboratory media. Under certain environmental conditions, some yeast strains may enter a sexual (or "perfect") phase in their life cycle. As part of this phase, they produce intermediate sexual spores called ascospores, which upon germination yield, once again, vegetative budding yeast. Aside from the biological importance of the sexual phase to the yeast, demonstration of ascospores is critical to separation of the yeast in question.   Absence of or

inability to demonstrate the presence of ascospores during
the life cycle results in the yeast (in this case,
Brettanomyces) being lumped into one grouping whereas
success in demonstrating a sexual phase places the yeast
(*Dekkera*) into another.   In all other criteria (i.e.,
utilization of sugar, nitrogen, requirements for vitamin
supplementation, etc.), the yeasts may appear to be
identical.
        There may well be several reasons unrelated to the
fundamental identity of the yeast for not successfully
demonstrating a sexual phase.   These may include use of
inappropriate sporulation media, need for nutritional
augmentation, absence of compatible mating type,
temperature of incubation, etc.   In the case at hand,
Dekkera requires a sporulation medium that includes
augmentation with several vitamins. Required in micro- and
milligram amounts, these nutrients are not easily and
routinely supplied in most production-oriented
laboratories.   As a result, suspect isolates are often
reported as "Brett-like" or *Brettanomyces/Dekkera*.
However, if differences in the sensory impact between the
two organisms growing in wine (or juice) is, in fact,
linked to the presence or absence of the sexual phase, then
the effort necessary to demonstrate the property is
warranted.   Ilagan (*13*) notes that even under ideal
conditions, relatively poor sporulation (<1%) is observed.

Distribution and Ecology

The involvement of *Brettanomyces* sp. in wine spoilage has
been reported from all wine-producing areas of the world
(*12, 19, 22*). In South Africa, van der Walt and van Kerken
(*35*) report *Brettanomyces* to be a frequent isolate from
yeast-associated spoilage (turbidity) in table wines.
        During the 1971 vintage, New Zealand researchers
reported isolation of *Brettanomyces* from 10 of 15 wineries
surveyed (*40*).   Unexpectedly, these workers reported more
frequent isolation from white than from red wine
fermentations.
        Compared with the report from New Zealand,
Brettanomyces infections in California are generally
observed in aging red wine, although infections have been
observed in Chardonnay and Sauvignon Blanc.   Schanderl and
Draczynski (*25*) and subsequently Van de Water (*29*) report
isolation of *Brettanomyces/Dekkera* from methode champenoise
sparkling wine en tirage.   Van de Water further notes that
Brettanomyces/Dekkera appears to be less sensitive than
Saccharomyces to carbon dioxide and concludes that they may
become a more widespread problem as California sparkling
wine production increases.
        Van der Walt (*30*) reports isolation of *Brettanomyces*
from honey and tree exudates.   It appears likely that
subsequent transmission to fermenting juice and wine is the
result of insect vectors.   Aside from this report, the
generally accepted habitat of both *Brettanomyces* and

Dekkera is from fermenting products and their associated environs.   It is likely that this ecological restriction is the result of their rather fastidious nutritional requirements.

It is clear from the world wine literature that attempts to isolate *Brettanomyces* and *Dekkera* from fruit in the vineyard have largely been unsuccessful. At least two reasons can be adduced for this failure. (1) Genera are clearly fastidious, requiring reasonably complex sources of exogenous nutrient including vitamin supplementation.   In clean fruit, proliferation would be very limited.   (2) Attempts to isolate minority populations from mixed flora are often frustrated by the presence of numerically superior and/or more rapidly growing species.

Once introduced into the winery, substantial populations can build up in difficult-to-clean sites such as equipment and transfer lines and valves where organic deposits may accumulate over the course of a season. Other important reservoirs for the organisms include drains and isolated pockets of juice and wine as well as pomace piles in close proximity to the winery. As evidence, Wright and Parle (*40*) report *B. intermedius* and *B. schnaderlii* from 25% of samples originating from such sites.

Spread of *Brettanomyces*/*Dekkera* populations within the winery can be attributed to use of contaminated and improperly sanitized equipment (pumps, hoses, etc.) and cooperage as well as insects common to fermentation facilities during the harvest season.   Principal among these is the common fruit-fly *Drosophila melanogaster*. Yeasts of all species represent an important food source for fruit fly adults and their larvae.   Since yeasts are unable to survive passage through the gastro-intestinal tract of adult *Drosophila*, the mechanism of dispersal is passive adherence to tarsi and other body surfaces of adult fruit flies during foraging (*26*).   Not coincidently, *Brettanomyces* can be recovered from winery locales that support growth of microorganisms and where fruit flies forage and breed.   Subsequent movement of adults around the winery potentially results in dissemination of large numbers of yeasts.

The most frequently cited locale for *Brettanomyces* and *Dekkera* within the winery is wood cooperage.   In this regard, recent observation implicates new cooperage as a more likely site for isolation than previously used barrels.   One reason for this is that species of both *Brettanomyces* and *Dekkera* produce the enzyme beta-glucosidase, that attacks the disaccharide cellobiose, producing glucose (*1*).   Cellobiose is an intermediate resulting, in part, from the charring or "toasting" process required to bend staves in the production of barrels.   However, cellobiose utilization varies between the species of *Brettanomyces* and *Dekkera* known to occur in grape wine. *B. intermedius* and *B. custersii* ferment and assimilate cellobiose whereas *B. lambicus* and *B. bruxellensis* cannot utilize the sugar

(34).   *Dekkera   intermedia*   is   able   to   utilize   the
disaccharide fermentatively and assimilatively whereas *D.
bruxellensis* cannot (33).   Cellobiose utilization is not
found in the wine-associated *Saccharomyces cerevisiae* (32).
  The   physical   properties   of   wood   also   contribute
significantly to microbiological control problems.   Unlike
polished stainless steel or glass, the inside surface of a
barrel presents a difficult-to-clean, irregular surface of
cracks   and   crevices   into   which   particulates   (including
spoilage yeasts) can settle.   Thus, organisms that grow in
wine (and fermenting juice) and happen to localize in such
an environment find themselves bathed in nutrient and
relatively protected against antagonistic environmental
pressures.   Their   apparent   resistance   to   the   common
preservative sulfur dioxide is one example of this.   Both
yeasts are, in fact, relatively sensitive to the effects of
molecular   sulfur   dioxide   (at   levels   of   0.8   mg/L)   when
suspended   within   the   volume   of   wine.   However,   their
frequent habitat between and deep within the cracks in
staves   often   protects   them   from   exposure   to   the
preservative.   Hence, substantially higher concentrations
of   sulfur dioxide might well be ineffective in control of
these sequestered populations.   Precipitated tartrates and
other fermentation debris present an even more impervious
barrier.
  During the course of a harvest season, it is entirely
possible   that   substantial   populations   of
Brettanomyces/Dekkera may become established in winery
cooperage, especially in facilities utilizing extensive
barrel fermentation.   Both *Brettanomyces* and *Dekkera* grow
relatively well in fermenting juice, and are capable of
attaining substantial population densities (albeit more
slowly than *Saccharomyces*).   In the laboratory we have
observed population increases of two orders of magnitude
over a one month period (*10*).   Aside from the role of
insect vectors in originating infection, another common
vehicle is acquisition of wine for use in blending or
topping material.   Any wine purchased from outside the
facility should be quarantined and not used until sterile
filtered (0.45 micrometer) or, at least, screened for
Brettanomyces/Dekkera.   In the case of established
populations, efforts should be made to identify and
eliminate pockets of high density populations that can be
transferred to uninfected wine.   A continuing program of
surveillance and sanitation is critical in excluding
"Brett"   from   the   winery   or   managing   established
infections.
  Once established in wood cooperage, elimination is
difficult.   Generally, multiple and thorough washes in hot
(>150°F) water appear only to be effective in short term
reduction in populations.   It is not known if recurring
growth results from failure to kill or subsequent
recontamination.   The practice of steaming the inside of
cooperage to 212°F, while seemingly effective in killing
contaminants, is potentially damaging to cooperage.

Because both *Brettanomyces* and *Dekkera* are slow growers and do not characteristically form a film or produce visually apparent quantities of carbon dioxide, their presence in barrel-aging wines may easily go undetected. Therefore, monitoring programs are essential for early detection and charting population changes. Although detection/monitoring programs expectedly vary between wineries, the method of choice generally involves collection of microbes by membrane filtration of known volumes of wine(s) and subsequent culture using differential media designed to identify acid-producing, actidione-resistant yeasts. When isolated from wine (and previously discussed caveats not withstanding), colony development after 6-7 days is generally regarded as confirmation for *Brettanomyces/Dekkera* (Smith, personal communication, 1993).

The time frame for monitoring begins at barreling, with subsequent examinations at racking and topping as well as prior to preparation of final blends and bottling. Since both yeasts are sensitive to sulfur dioxide, it is recommended that laboratory personnel collect samples for plating prior to sulfiting operations.

Due to the labor-intensive nature of monitoring programs, alternatives have been sought. Enzyme-Linked Immunosorbent Assay (ELISA) has been proposed (*14*). The advantage of ELISA is that viable yeasts need not be present. Unfortunately, the method is **too sensitive** for routine production applications and, at present, too costly.

Brettanomyces in barrel-aging wines follows a bell-shaped growth pattern, reaching maximum population density 5-7 months after vinification. The time frame for development of maximum cell number (and subsequent decline) depends, in large part, on wine chemistry (particularly levels of molecular sulfur dioxide and available fermentable sugars) as well as cellar temperature.

Until recently, containment/elimination of *Brettanomyces* utilized sulfur dioxide additions at rackings or when populations were observed to increase. With increasing concern regarding levels of sulfur dioxide, winemakers have begun to consider alternate strategies, opting for "management" of microbiological problems. Lowering cellar temperature to <55°F is one cellar management technique known to be useful in slowing and potentially preventing growth of *Brettanomyces/Dekkera* (*27*). Other long-term goals may well include identification and utilization of potential antagonistic relationships between microorganisms that deprives "Brett" of the opportunity to grow. Further, "sensory-neutral" strains of both *Brettanomyces* and *Dekkera* may well exist. In practice, such strains could be used to bring about conversion of the same substrates as offensive strains, but without producing the deleterious metabolites.

While the "school-of-thought" for dealing with "Brett" has changed somewhat from complete elimination to management of established populations, winemakers generally agree that it is necessary to bottle wine with no, or very low numbers, of viable cells. In this regard, options vary with available winemaking technology and philosophy. Assuming the winemaker doesn't wish to sterile filter, this likely means that wines will not be bottled until resident populations decline. The point at which a winemaker feels "safe" with respect to bottling depends upon cell numbers (and whether the population appears to be increasing or declining), wine chemistry, and previous history with their particular "house strain." Some believe that availability of seemingly minute amounts of unfermented sugar predisposes wines to growth of *Brettanomyces* once they are bottled. While our own research supports the fact that growth rate is enhanced with increasing concentrations of glucose, substantial populations of both *Brettanomyces* and *Dekkera* may develop at levels of less than 0.2%. Smith (27) estimates that (theoretically) hexose concentrations of 100 mg/L can support *Brett.* populations of 10^7 CFU. Further, populations do not necessarily need to be increasing to produce undesirable sensory effects. Thus the fact that wine is considered "dry" (by generally accepted standards) at bottling doesn't, by itself, mean that *Brettanomyces* will not develop.

Are there other substrates present in wine and juice that may be available to *Brett.* as sources of carbon and nitrogen? Ethanol and ethyl acetate are assimilated aerobically and may represent sole carbon sources. Smith (27) reports that D-proline represents a sole nitrogen source. In that this amino acid is not biologically available to *Saccharomyces* during the course of fermentation, it is normally present in wines at substantial levels (16).

Sterile filtration represents an efficient and effective means for removal of microorganisms prior to bottling. However, winemakers do not universally agree as to the benefits versus risks of membrane filtration. While everyone can agree on the benefits of bottling a wine free of troublesome yeasts and bacteria, many still feel that filtration of red wines through a 0.45 micrometer membrane compromises and diminishes the sensory impact of the product. Although we disagree, it is clear that filtration of any wine that has supported a dense (>10^4 CFU) population of *Brettanomyces* or *Dekkera* will not alleviate the objectionable sensory profile. Properly used, sterile filtration is effective in preventing future activity in the bottle.

Another tool has recently become available to the winemaker for dealing with yeasts and bacteria present at bottling. Dimethyldicarbonate (DMDC), marketed under the tradename "Velcorin," has been shown to be very effective in killing both wine yeasts and bacteria (20). However, special equipment is required for incorporation.

Winemakers also agree that *Brettanomyces* activity, if unavoidable, must be restricted to barrel-aging wines.

In this environment, metabolites (in low concentration) potentially add complexity and can be dealt with during formulation of blends prior to bottling. The extent to which yeast growth adds to complexity or results in diminished character depends on cell numbers and residence time in the wine. Generally speaking, higher cell numbers and longer the contact times results in greater sensory impact. Since growth is stimulated by fermentable sugars, levels at barrelling should be as low as possible (<1 g/L).

As discussed in our introductory remarks, various sensory descriptors have been used to characterize *Brettanomyces/Dekkera*-tainted wines. These range from "spicey" to "mousey" in character. The origins of spicey and mousey components have been reported in the literature. Heresztyn (*12*) reports that fermentations utilizing *B. intermedius* and *B.anomalus* produced substantial amounts of the volatile phenols 4-ethyl guaiacol (arising from ferulic acid) and 4-ethyl phenol (from p-coumaric acid). The former is characterized as being clove or spice-like whereas the latter is described as smokey or medicinal. Sinapic acid was also labile, yielding 4-ethyl syringol and 4-vinyl syringol in varying amounts. However, Heresztyn reported that the latter two volatile phenols have minimal sensory properties compared to others produced. 4-Ethyl guaiacol is variously described as having a strong clove to smokey character which appears to be dependent upon its concentration and the matrix in which it is formed (*8*).

Among the volatile phenol compounds identified, 4-ethyl phenol (produced from p-coumaric acid) was reported to be present in highest concentration. This suggests its utility as a general sensory marker in *Brettanomyces*-infected wines.

It is believed that these compounds, characteristic of *Brettanomyces* and *Dekkera* metabolism, result from decarboxylation of hydroxycinnamic acids yielding vinyl phenol intermediates and subsequent reduction to produce the ethyl analog (*28*). As seen in Figure 1, initial decarboxylation is mediated by a substituted cinnamate decarboxylase (SCD) while the reduction step utilizes a Vinyl-Phenol Reductase (VPR).

Dubourdieu (*7*) points out that wine yeasts *Saccharomyces cerevisiae* also contain the cinnamate reductase, and thus are capable of producing the vinyl phenol intermediate. However, flavonoid phenols (tannins) inhibit its activity and, hence, formation of volatile phenols in red and rose wines is significantly less than is seen in white wine fermentations. Activity of SCD in the case of *Brettanomyces* and *Dekkera*, however, is not inhibited by polymeric phenols.

Formation of "mousey" components associated with *Brettanomyces*-infected wines has also been studied (*2, 11, 12*). The offensive compounds are reported to be the ethyl amino acid derivatives, 2-acetyl-1,4,5,6-tetrahydropyridine and 2-acetyl-3,4,5,6-tetrahydropyridine. Synthesis by *B. intermedius* and *B. lambicus* requires lysine and ethanol.

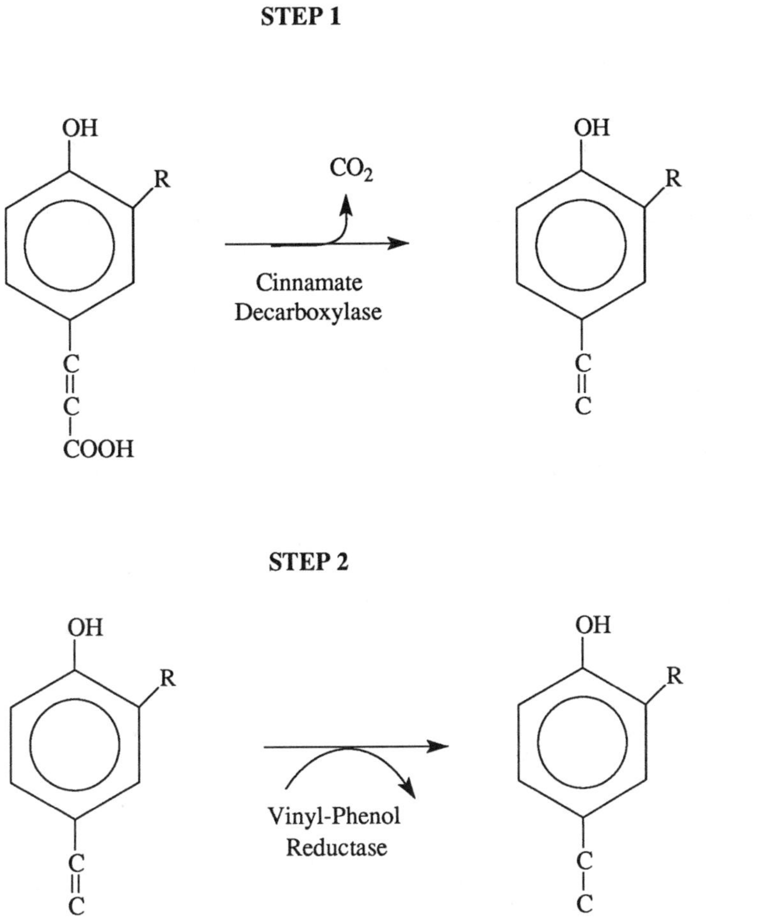

STEP 1

STEP 2

R = H : p-coumaric acid   R = OCH$_3$: 4-vinyl-guaiacol
R = CCH$_3$: ferulic acid   R = H : 4-ethyl-phenol
R = H : 4-vinyl-phenol   R = OCH$_3$: 4-ethyl-guaiacol

Figure 1.   Proposed Pathway for Formation of
Volatile Phenols by *Brettanomyces* and *Dekkera*
sp.

In this study, *Brettanomyces* isolates produced the mousey compounds when growing both in a fermentative mode or in "dry" wine.   It is interesting to note that the heterofermentative lactobacilli used in this study (*Lactobacillus intermedius* and *L. cellobiosus*) also produced the same metabolites from lysine and ethanol.

Work carried out at C.S.U. Fresno over a three year period has clearly shown that much of the final sensory properties in *Brettanomyces* and *Dekkera*-tainted wines result from presence of either organism during primary fermentation (*10*). Using sterilized (DMDC 350 mg/L) French Colombard juice and established cultures of *Saccharomyces cerevisiae* (UCD 522) as well as *Brettanomyces* (UCD 605) and *Dekkera* (UCD 615), fermentations were carried out in pure and co-culture at initial population densities of 10^6CFU/mL.   Changes in the relative populations were followed by daily microscopic examination (see Figure 2). Aliquots of each fermentation were also collected daily and frozen for later use.   Upon completion of co- and pure culture fermentations, each lot was clarified, sterile filtered (0.45 um), stabilized with 350 mg/L DMDC and 0.8 mg/L sulfur dioxide, and held at 5°C for further analysis.

Tastings   were   conducted   with   wine   industry representatives   and   CSUF   staff   at   one   month   post fermentation.   Pure culture *Brett*. fermentations were described as being reminiscent of "old cider".   This description is consistent with the literature.   Pure culture *Dekkera* fermentations were described as having a strawberry and peach-like character but with "sweaty" or "malty"   notes.   By   comparison,   the   pure-culture *Saccharomyces* control was described as being typical of the varietal.   None of the over 20 people that tasted these samples suggested the presence of typical descriptors of "horsey" or "mousey" for these wines at this stage of maturation.   It should be noted that each lot was examined by   HPLC for the presence of lactic acid, which would suggest LAB activity.   In the case of both *Brett*. and *Dekkera*, detectable levels of lactic acid were not found.

Tastings of co-culture lots were done in conjunction with pure culture lots.   Similar sensory discriptors (although of diminished intensity) were reported.   As can be   seen   in   Figure   2,   resident *Brett*. and *Dekkera* populations in co-culture underwent only one budding cycle during this short time frame.   Thus, although cell number did not approach that seen in pure culture after 20-25 days, a similar (although much decreased) sensory impact could definitely be detected.   It appears as though both strains are capable of bringing about significant sensory effects that are not directly linked to increased cell number.   In the case of mixed fermentations, some of the respondents indicated a preference for coded lots which were later found to be *Brett*. and *Dekkera* co-cultures. Other   members   of   the   group   did   not   voice   specific complaints except to note that some of the coded samples "had diminished fruit," a property consistent with growth

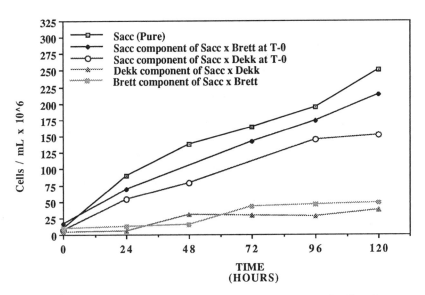

Figure 2.   Growth Curves for Pure- and Mixed-
Culture Fermentations.

of both yeasts.   The group could not consistently identify
treated vs. control wines at this early stage.
     Three months later, each lot was retasted.   At this
point, both *Brett.* and *Dekkera* pure culture lots were noted
to have developing "horsey" and "mousey" odors.   Mixed
fermentation lots had not changed appreciably from 1 month
tastings.
     At   seven   months   post   fermentation,   samples   of
co-culture   and   pure   culture   *Saccharomyces*   lots   were
comparatively tasted again by a group of 38 winemakers.   At
this point, differences between lots could be detected (p
= 0.05).   However, no clear preference was seen between
pure   culture   *Saccharomyces*   versus   co-culture
fermentations.   Among   those   participants   able   to
distinguish between pure culture *Saccharomyces* and mixed
culture   fermentations,   "diminished   fruit",   "increased
complexity" and "aged" were frequently recorded comments.
     We believe that it is important to note that in those
lots where the issue of complexity (positive) was raised,
exposure time to either *Brett.* or *Dekkera* was limited (4-5
days)   at   which   time,   cell   contact   was   terminated   and
samples stored under sterile conditions. Although untested,
we subscribe to the hypothesis that continued cell contact
(similar to pure culture fermentations) may have resulted
in the distinctly negative descriptors noted in those lots.
     Aside   from   their   own   unique   contributions   to   wine
character, the presence of either strain has been shown to
repress the activity of *Saccharomyces*.   As can be seen
from Figure 2, pure culture populations of *Saccharomyces*,

growing fermentatively, reach higher cell densities than are seen in fermentations where either *Brettanomyces* or *Dekkera* were added as a co-inoculum at the start of fermentation. From Figure 2, it can be seen that both *Brettanomyces* and *Dekkera* appear to undergo a single budding cycle over the course of fermentation. This is not surprising in that both grow relatively slowly. However, the fact that they were able to repress the growth of the *Saccharomyces* component of fermentation prompted further investigation.

Initially it was felt that growth suppression resulted from competition for space in the fermentation volume. Follow-up studies comparing growth profiles of initial *Saccharomyces* populations of 1×10^6 through 6×10^6 CFU revealed that growth rates did not significantly vary except at the highest initial cell inoculum. From this we concluded that, at least at cell titers used in this study, competition for space was not an issue. A second and more likely source of repression is formation of inhibitory fatty acids by the spoilage yeasts involved. As seen in Table 1, it is apparent that known inhibitory fatty acids (acetic, octanoic and decanoic acids) are present in much higher concentrations in co-culture fermentations compared with levels seen in pure culture *Saccharomyces* fermentation. To test whether levels observed resulted in growth suppression profiles similar to those seen in Figure 2, octa- and decanoic acids were added to a second set of pure culture *Saccharomyces* fermentations in amounts equivalent to highest levels reported in the co-culture fermentation. Resultant growth repression similar to that seen in Figure 2 for *Saccharomyces* populations was observed.

Conclusions

Species of *Brettanomyces* and *Dekkera* appear to be ubiquitous and difficult to control. Several reasons can be identified.

(1) Intra- and interstate sale of bulk wine has been and continues to be a thriving enterprise. Winemakers would be well advised to quarantine all such shipments upon receipt, until thorough microbiological examination has been conducted to ascertain that no contamination exists. As a "safety-net" we strongly encourage sterile filtration of all wine coming on the premises.

(2) Since both yeasts are frequent isolates from cooperage, purchase of used barrels/wood tanks should be considered as risky. At the very least, wines produced using suspect cooperage should be kept separate from the rest of the winery's inventory until they have been demonstrated to be free of infection.

(3) The unobtrusive growth pattern of both yeasts permits substantial populations to become established before the problem is recognized. Reliance on sensory monitoring rather than regular microbiological sampling

will result in more extensive dissemination within the winery.

(4) Current practices of using little (or no) sulfur dioxide in winemaking have led to microbiological control problems. Although no single substitute has yet to be reported, efforts to identify alternatives, including identification and exploitation of antagonistic microbiological interactions, need to be intensified.

Over the last ten years, winemaking posture with regard to *Brettanomyces/Dekkera*, has shifted from denial ("It doesn't exist in my winery!") to one of complete control. Today, most winemakers speak in terms of "management" of established populations. In some instances, wineries are evaluating the potential for controlled utilization of *Brettanomyces* and/or *Dekkera* in varietal wine production. We have pointed to instances in our own studies where aroma/bouquet and flavor modification arising from **controlled** co-culture fermentations has been viewed positively by winemakers. Especially important are observations of "diminished fruit" and "enhanced complexity" as well as "aged character" observed in very young (<2 months) wines in which either *Brettanomyces* or *Dekkera* were present in co-culture with *Saccharomyces* for short periods of time during fermentation. These may represent sought-after attributes by some winemakers. **However,** we want to emphasize that intentional use of these poorly understood genera as stylistic tools in winemaking is still a dangerous practice.

To our knowledge, reports of so-called "benign" strains have not been documented. At present, we have one such strain in culture, and plan to perform comparative studies with it against certified strains.

Acknowledgments

We would like the thank the American Vineyard Foundation and the California Agricultural Technology Institute for financial support of this project. We would also like to thank Mr. Clark Smith (WineSmith, Napa, CA 94558) and Ms. M. Bannister (VINQUIRY, Healdsburg, CA 95448) for their critical reviews.

Literature Cited

(1) Blondin, B.; Ratomahenina, R.; Arnaud, A.; Galzy, P. *Biotech. and Bioengineer*; John Wiley and Sons: New York, NY, 1982, Vol. XXIV, pp 2031-37.

(2) Craig, J.T.; Heresztyn, T. *Am. J. Enol. and Vitic.* **1984**, *35(1)*, 46-47.

(3) Davis, C.R.; Wibow, D.; Eschenbruch, R.; Lee, T.H.; Fleet, G.H.   *Am. J. Enol. and Vitic.* **1985**, *36*, 290-301.

(4) Davis, C.R.; Wibowo, D.; Fleet, G.H.; Lee, T.H. *Am. J. Enol. and Vitic.* **1988**, *39(2)*, 137-142.

(5)   Drysdale, G.S.; Fleet, G.H. *Am. J. Enol. and Vitic.*
      1988, *39(2)*, 143-54.
(6)   Drysdale, G.S.; Fleet, G.H. *Am. J. Enol. and Vitic.*
      1989, *40(2)*, 99-105.
(7)   Dubourdieu, D.   Influence of Yeasts on Some Grape
      Aroma Precursors. Presented at Recent Advances in
      Oenology, Fresno, CA, 1992.
(8)   Dubois, P. In *Flavour of Distilled Beverages: Origin
      and Development;* Piggott, J.R., Ed.; Soc. of Chem.
      Ind.: London, 1983, pp 110-19.
(9)   Fugelsang, K.C.; Zoecklein, B.W.   *Pract. Winery and
      Vineyard* **1993,** May-June.
(10) Fugelsang, K.C.; Osborn, M.M.; Muller, C.J.   Proc. of
      the Wine Ind. Tech. Symp.; Rohnert Park, CA, 1992.
(11) Heresztyn, T. *Arch. Microbiol.* **1986,** *146*, 96-98.
(12) Heresztyn, T.   *Am. J. Enol. and Vitic.* **1986,** *37(2)*,
      127-132.
(13) Ilagan, R.D.   *Studies on the Sporulation of Dekkera;*
      M.S. Dissertation; University of California: Davis,
      CA, 1979.
(14) Kuniyuki, A.H.; Rous, C.; Sanderson, J.L.   *Am. J.
      Enol. and Vitic.* **1984,** *35(3)*, 143-45.
(15) Kunkee, R.E.; Amerine, M.A.   In *The Yeasts;* Rose,
      A.H.; Harrison, J.S., Eds.; Academic Press: London and
      New York, 1970, Vol 3, pp 6-60.
(16) Ough, C.S. *Vitis* **1968,** 7, 321-31.
(17) Pardo, I.; Garcia, J.J.; Zungia, M.; Uruburu, F.
      Appl. Environ. Microbiol. **1989,** *55*, 539-541.
(18) Peynaud, E. *Knowing and Making Wine;* John Wiley and
      Sons: New York, NY, 1984.
(19) Peynaud E.; Domercq, S.   *Am. J. Enol. and Vitic.*
      1959, *10*, 69-77.
(20) Porter, L.J.; Ough, C.S.   *Am. J. Enol. and Vitic.*
      1982, *33(4)*, 222-225.
(21) Rankine, B.C. *Aus. Wine Brew. Spirits Rev.* **1963,** *81*,
      11-16.
(22) Rankine, B.C.; Pilone, D.A.   *Aust. Wine Brew. Spirits
      Rev.* **1974,** *92*, 36-40.
(23) Reed, G.; Nagodawithana, T.L.   *Yeast Technology;* Van
      Nostrand Reinhold Publ.: New York, NY, 1991.
(24) Sanfacon, R.; Roulliard, R.; Heick, H.M.C.   *Can. J.
      Microbiol.* **1976,** *22*, 213-220.
(25) Schanderl, H.; Draczynski, M.   *Dt. Weinztg.* **1952,** *88*,
      462-64.
(26) Shehata, A.M.; Mrak, E.M.   *Amer. Naturalist* **1951,** *85*,
      381-83.
(27) Smith, C.   *Studies on Sulfur Dioxide Toxicity for Two
      Wine Yeasts;* M.S. Dissertation; University of
      California: Davis, CA, 1993.
(28) Steinke, R.D.; Paulson, M.C.   *J. Agric. Food Chem*
      1964, *12*, 381-87.
(29) van de Water, L. quoted in: *Coping with Brettanomyces*
      by S. Hock. *Pract. Winery and Vineyard* **1990,** Jan/Feb.,
      pp 26-31.

(30) van der Walt, J.P.   In *The Yeasts, A Taxonomic Study*; Lodder, J., Ed.; North Holland Publ. Co.: Amsterdam, London, 1970, 2nd ed., pp 863-92.

(31) van der Walt, J.P.   In *The Yeasts, A Taxonomic Study*; Lodder, J., Ed.; North Holland Publ. Co.: Amsterdam, London, 1970, 2nd ed., pp 157-165.

(32) van der Walt, J.P.   In *The Yeasts, A Taxonomic Study*; Lodder, J., Ed.; North Holland Publ. Co.: Amsterdam, London, 1970, 2nd ed., pp 555-718.

(33) van der Walt, J.P.   In *The Yeasts, A Taxonomic Study*; Kreger-van Rij, N.J.W., Ed.; Elsevier Science Publishers B.V.: Amsterdam, 1984, 3rd ed., pp 146-150.

(34) van der Walt, J.P.   In *The Yeasts, A Taxonomic Study*; Kreger-van Rij, N.J.W., Ed.; Elsevier Science Publishers B.V.: Amsterdam, 1984, 3rd ed., pp 562-576.

(35) van der Walt, J.P.; van Kerken, A.E.   *Antonie van Leeuwenhoek*, **1961**, *27*, 81-90.

(36) van Wyk, H.   *Am. Chem. Soc. Abstr.* **1973**, *165*, 59.

(37) Wang, L.F.   *Off Flavor Development in White Wine by Brettanomyces and Dekkera*; M.S. Dissertation; California State University: Fresno, CA, 1985.

(38) Watanakij, P.   *Volatile Fatty Acid Composition of Wines Fermented by Brettanomyces and Dekkera Yeasts*; M.S. Dissertation; California State University: Fresno, CA, 1992.

(39) Wayman, M.; Parekh, R.S.; Parakh, S.R.   *Biotech. Letters* **1987**, *9(6)*, 435-40.

(40) Wright, J.M.; Parle, J.N.   *New Zealand J. of Agric. Res.* **1974**, *17*, 273-78.

(41) Wyman, C.E.; Spindler, D.D.; Grohman, K.; Lastick, S.M.   *Proc. Biotech. Bioeng. Symp.* **1986**, *17*, 221-238.

(42) Zoecklein, B.W.; Fugelsang, K.C.; Gump, B.H.; Nury, F.S.   *Production Wine Analysis*; Van Nostrand Rheinhold: New York, NY, 1990.

RECEIVED March 7, 1993

TECHNOLOGICAL APPLICATIONS IN PRODUCTION

Chapter 8

Applications of Technology in Wine Production

Richard P. Vine

Department of Food Sciences, Purdue University, West Lafayette, IN 47907

The more recently created body of scientific wine knowledge supports a current technology that is significantly different from the conventional wisdom employed by vintners even less than a decade ago. Rationale now exists to justify new methods in virtually every phase of wine production.

This chapter will consider some of the more important developments that have surfaced in the enology discipline and are currently applied in the wine industry.

Optimal Maturity of Grapes

Some commercial winemakers continue to monitor soluble solids, in the form of Brix measurement, as the principal grape ripeness indicator. Others have adopted one of several formulae in which pH and total titratable acidity analyses are factored in a ratio devised to predict optimal maturity.

Robredo (8) conducted a study in which data gathered from these traditional harvest analyses were taken in tandem with finite HPLC measurements of important flavor compounds. The objective was to assemble a biochemical model that could be used to establish more exacting parameters in declaring ideal grape maturation. The methodology was to quantify individual sugars and organic acids in grapes which influence Brix, T.A., and pH, along with specific ester precursors and phenols. One white and one red variety were analyzed at different times during the ripening process and results were correlated with wines that achieved superior sensory scores. The conclusion was that traditional analyses indicated an optimal 'industrial' harvest date for the white variety at least two days later than the optimal 'technical' harvest

0097–6156/93/0536–0132$06.00/0

time determined from the new analytical profile. The optimal industrial and technical harvest dates for the red variety were the same.

The practical application of this technology remains for the future. HPLC requirements for this type of control are currently out of reach by most vintners - although technology is also rapidly closing this gap. Some universities and other institutions may be able to offer interim assistance to vintners interested in this new element of quality assurance. More precise sampling techniques in the vineyard will also be needed to ensure accurate predictions.

In the meantime, there also remains the need for further research to determine optimal organic constituent profiles for other grape varieties in various environs. Research in this vein may also identify specific constituency relationships which can enhance certain production parameters, such as maximizing stability of color pigmentation and protein solubility.

Acceleration of Fermentation and Fining in Late-Harvest Musts

Despite the lack of wide market appeal, vintners respond to the connoisseur niche by offering an increasing number of late-harvest wines in U.S. markets.

Some of these wines are made from grapes having naturally raisinized by excessive sun exposure in the vineyard. Sluggish fermentation rates in higher Brix musts resulting from late-harvest grapes can be attributed to an increase in water activity - an osmotic force taking up water which places a dehydration strain called zytorrhysis on yeast cells. An increase from 20 to 50° Brix can reduce yeast cell volume by about 50%. Minimizing Brix levels, and therefore reducing the impact of this phenomenon, has brought closer vintner scrutiny upon identifying a rather precise amount of ethanol, residual sweetness, and acid balance desired to meet perceived consumer demand.

Amidst cooler temperatures and higher humidity *Botrytis cinerea*, the 'noble mold', can appear as fuzzy gray spots on grape skins during the harvest season. Botrytis spores bore through the skins and effect an evaporation of berry water - intensifying retained sugars, acidity, glycerol, and flavors. The infection process actually reduces sugar and acidity (increases glycerol and mucic acid), but the net effect is a concentration. Botrytized fruit flavors undergo change due to the formation of ethyl esters of hydroxy-, keto-, and dicarboxylic acids. Yeast enzyme synthesis and activity rates determine how these acids are metabolized into the ultimate ester flavor profile. Some natural fruit flavors, such as the aromatic monoterpenes, geraniol and linalool, common to the Johannisberg Riesling, Gewurztraminer and Muscat varieties, are destroyed by botrytization (*15*).

The mold can also form laccase, an enzyme capable of oxidizing important phenols, including anthocyanins. This accounts for the golden-brown colors in most botrytized wines. Botrytis infection can be quantified using a laccase assay available in laboratory kits. Samples are initially treated with PVPP to reduce polyphenol interference and then mixed in a spectrophotometer cell with a syringaldazine-ethanol solution, along with a buffer. The change in absorbance is recorded over several minutes and calculated by formula to provide laccase activity measured in laccase units per mL.

The botrytis fungus can reduce up to half of the protein, amino acid, and free ammonium nitrogen (FAN) constituency of grapes. Constrained FAN can induce deamination activity by yeasts upon protein and amino acid constituents - causing the development of unpleasant hydrocarbon flavors and other maladies of flavor. FAN is a more essential element in active yeast growth and deficiencies can be supplemented with diammonium phosphate and other food-grade sources.

A more serious problem is a reduction of thiamine and other B complexes necessary for decarboxylation in the pyruvic acid cycle and the synthesis of various keto acids mentioned above. These vitamins can also be supplemented - often as components in concert with FAN in commercial proprietary mixtures such as Yeastex [R]. Yet another concern is existing evidence indicating botrytis activity can produce trace amounts of antibiotics that may be toxic to yeasts (15).

Botrytis infection produces polygalacturonase which hydrolyses pectins into polysaccharides. Mucic acid development can react with calcium to form slow-developing precipitates. Both are hurdles to effective fining - evidenced by frequent observation of hazy wines in the bottle. Proper use of glucose oxidase enzymes followed by appropriate applications of kieselsol fining mentioned below can provide additional help in overcoming problem clarifications.

Moderation of these perils and pitfalls remains best controlled by close monitoring of botrytis development in the vineyard and blending techniques in the cellar.

Control of Natural Microorganisms

It remains rather commonplace for vintners to add 30-90 mg/L of free sulfur dioxide to grapes in the crusher, or early on in the resulting must.

The traditional rationale for sulfite additions at the crush is that this dosage inhibits or kills wild yeasts, with a more secondary control of bacteria and molds, as well as some protection from oxidation. Research findings from Panagiotakopoulou and Morris (6) indicates that SO_2 additions at the crusher actually increase browning in resulting white wines. Traditional problems with this technique are that high pH grape musts reduce molecular SO_2 and, therefore, its effectiveness; poor and/or extended storage conditions of potassium metabisulfite and other SO_2 sources reduce ion availability; deficient dosage calculations leaving the must unprotected; as well as excessive dosage calculations inhibiting cultured yeast and bacteria inoculations.

While 'killer' yeasts are not new to enology, their role in reducing or eliminating SO_2 at the crusher is an application that is relatively recent. Killer yeasts are species (the original isolates were *Saccharomyces cerevisiae*) that kill sensitive members of their own species and frequently those of other species, as well. They function by secreting a plasmid-coded protein toxin that binds with 1,6 beta D-glucan receptor components in the cell walls of sensitive strains. This toxin interacts directly with protein components of the cell membrane and, in turn, disrupts the normal state of cell activity. Boone et al. (1) suggest that killer yeasts are immune due to a precursor protein that functions as an inhibitor of toxin in its cell membrane metabolic processes.

Killer yeasts are particularly effective in reducing infection from *Brettanomyces*. This spoilage microorganism is most often identified with red wine spoilage in the form of acetic, isobutyric and isovaleric acids which emerge as pungent, 'mousey' or 'horsey' odors. *Brettanomyces* can resist mid-range dosages of free SO_2, is insensitive to sorbic acid, and may go unnoticed until growth has become widespread. Consequently, musts undosed with SO_2 require immediate inoculation with killer yeast cultures in order to achieve maximal protection. Some enologists inoculate at the crusher hopper - taking advantage of heavy oxygen demand by the yeasts to lower the oxidation potential.

Van Vuuren and Jacobs (10) report that the killer system occurs in some natural yeast strains. Musts inoculated with a sensitive yeast strain culture can be dominated by wild killer strain populations, causing stuck fermentations. These researchers point out that the resulting wine can suffer from reduced ethanol yield, high volatile acidity, formation of H_2S and contaminant flavors caused by acetaldehyde, fusel oils, and lactic acid.

Vintners continue to evaluate a growing availability of killer yeast strains. Popular strains of killer yeasts are <u>Champagne 111</u> and <u>Montrachet 1107</u>, the latter more popularly known as 'Prisse de Mousse'. These and other strains of cultured killer yeasts are frequently used by progressive winemakers.

Control of Oxidation

Sims et al. (9) report that, while non-sulfited musts contain high levels of polyphenoloxidase and resultant pigmentation oxidation, the delay of sulfiting ultimately results in wines of reduced total phenolics and improved sensory quality. Panagiotakopoulou and Morris (6) conclude that appropriate additions of ascorbic acid can aid in minimizing persistent browning due to oxidation.

Free oxygen is rapidly utilized by cultured *Saccharomyces* spp. wine yeasts (facultative anaerobes) at the outset of logarithmic growth - with full depletion of O_2 resulting in anaerobic fermentation. This condition results in the redistribution of certain lipid and sterol compounds essential for yeast membrane construction during cell division - and membrane mechanics during the glycolytic functions of fermentation. Some winemakers actually supplement oxygen by agitating white wine fermentations when activity commences to diminish - more often in combination with FAN supplements, as well.

Oxygen present during the post-fermentation processing of young white wines requires an entirely different approach. The greater a wine's buffering capacity, or its content of oxidizable compounds, the greater its aging potential. Aging is thus a controlled process of oxidation. Consequently, wines containing higher levels of yeast autolysis compounds (wine fermented sur lies, or treated with significant additions of yeast hulls) and/or containing a bit of free SO_2, are chemically equipped to bind with greater quantities of oxygen. The result is often measured in wines having less browning, with more complexity and structure in the flavor profile. Higher levels of phenols, as found in longer-term skin-contact musts, also influence a higher buffering capacity, and account for the major reason why red wines, particularly at lower pH levels, generally take far longer periods of time in aging to maturity.

While oxygen can be employed effectively in making complex white table wines, those designed to be lighter in style, more fresh and fruity in aroma, should be made from generally lower pH grapes afforded minimized oxygen exposure. Sulfiting at the crusher may continue to be advisable as resulting increases in phenols are often balanced with residual sweetness in these types of wine. Cold temperature short term aging in stainless steel tankage, nitrogen sparging, and adequate maintenance of free SO_2 levels are essential for optimal quality in these wines.

Red Wine Color Enhancement and Stability

Some red grape varieties, such as Pinot Noir, are often color deficient - certainly in comparison to the dense pigmentation generally found in wines made from Cabernet Sauvignon and Petite Sirah. While some wine aficionados accept delicate color values as part of varietal character, indeed, darker Pinots are held suspect by wine judges, some question remains whether or not the overall image of red wines is negatively influenced by modest hue intensities.

Sugar residues in the formation of anthocyanin types,

i.e., monoglucosides and diglucosides, play a dramatic role in wine color stability, as do the types of anthocyanins themselves. Monoglucoside forms of malvadin and peonidin are dominant in most commercially-grown red varieties of *Vitis vinifera*. Diglucoside forms of delphinidin and petunidin, found in *Vitis labrusca* and other species, are the least stable. Hybrid cultivars, as would be expected, exhibit a wide range of color hue, intensity, and stability in relationship to parental genetic influence. Optimal color stability can be achieved by closely monitoring pH during the harvest season. Lower pH, in the 3.20-3.30 range, is generally associated with richer, more purplish tones, while higher pH ranges are usually expressed in brick-ruby hues.

Vintners often choose to enhance color by separating free-run or lightly pressed juice for pink 'blush' wines or 'blanc de noirs' table and sparkling wine cuvees. The remaining pomace is then added to other crushed grapes to increase the availability for increased pigment extraction when fermented in traditional skin contact methods. Exacting control over the extent of extraction is necessary in order to avoid generating distorted flavor profiles and saturated concentrations of phenolics - as well as excessive astringency, bitterness, and eventual color precipitates.

A more simple method is by simple blending of a 'teinturier', a dense, inky wine made from Salvador, Colobel, and other heavily pigmented grape varieties. Sometimes heavily-pigmented press wine fractions are employed in a similar manner. Vintners differ in their approach to this as some feel the blend strays from the ideals of varietal purity, while others point out that many classic reds, such as Bordeaux, are blends among different varieties anyway.

New products made possible by membrane separation processing techniques explained below have resulted in the isolation of concentrated pigments in retentate form. While this may carry similar concerns of varietal purity, it does so on a far smaller scale as comparatively little pigment is required to achieve favorable results. A product called Xpress[R] is now available in several forms designed to fit the most common needs in color enhancement.

Reduction of Astringency and Bitterness

The extraction of complex polyphenols from grape seeds, skins, and stems, has been a problem of varying magnitude. Commercial white wines generally have only minor constituencies of phenols due to the extraction of juice prior to fermentation. On the other hand, excessive treatments in the crusher and press (generally to maximize juice yield) create the risk of cracked seeds, macerated skins and stem fragments contributing significantly to astringency and bitterness in the finished wine. In red

wines the conventional vinification method is to ferment
the must in total contact with grape solids in order to
release anthocyanin color pigments from the skins - a
process ranging from several days up to several months.
This technique aggravates extraction of astringent and
bitter phenols - conditions often referred to as 'tannic'
and 'harsh' in wine jargon.

Most enologists categorize phenols into two major
groups. Polymeric flavonoids comprise the largest
fraction of phenolic compounds in wines and can be traced
to the processing described above, as well as from
degradation of larger molecular components. Flavonoids
serve as oxygen reservoirs which contribute to oxidation
reactions important in wine aging and development. Non-
flavonoids comprise a much smaller portion of phenol
constituency in wines and are generally more aromatic,
particularly the aldehyde compounds resulting from wood
aging regimens. With this diversity in attributes, close
attention to the management of overall phenolic profile in
wines is an essential element in wine quality.

Up until relatively recently, grape crusher-stemming
machines have been a major source of excessive phenols.
Some older devices served to mill the grapes into small
particles of solids creating an immense surface area for
extraction. Later models were equipped with adjustable
rollers in order to reduce maceration.

Contemporary units such as the AMOS [R] are available
that separate grape berries from the stem rachis without
crushing. Some winemakers have removed crusher rollers
altogether and dump their grapes directly into the
conventional destemming chamber. Stem tannin extraction
can then be more precisely monitored by adding back a
desired percentage of stems, if any, to the must. The
application of these principles have had a very positive
impact upon reducing wine astringency and bitterness.

Similar history has evolved in press equipment.
Early mechanization for rotating the press basket served
to increase juice yields by the movement of must passing
against the screen walls. These dynamics also increased
commensurately more phenolics. The pressing of red wines
after must fermentation is particularly sensitive to
phenol extraction - with most winemakers insisting upon
the separation of 'free run' and 'press' wines. Later
evaluation of the press wine determines its percentage in
the assemblage of the final wine blend (13).

The tank press is comparatively new to commercial
winemaking, and better methods of application have
continued efficiency in yields while markedly lowering
phenolic extraction. The device consists of a closed
stainless steel horizontal tank in which a membrane is
constructed so as to divide the interior in half
lengthwise. The membrane separates the tank into a press
chamber and a pneumatic chamber. Inflation of the latter
exerts a firm but gentle and even pressure upon the must

on the opposite side of the membrane. There are no baffles, chains, rings, or other devices employed to loosen the must/pomace cake - a major source of maceration in the older rotating basket presses. Tank presses offer the distinct advantage of resisting pomace cake formation due to the elasticity of the membrane. Figure 1 illustrates the liquid extraction process due to the slowly moving must against the channel outlet pores situated upon the inner press chamber. This creates a far more gentle agitation as the tank rotates - and significantly reduces phenolic concentrations. Some winemakers have achieved red wine quality so high that they no longer have the need to separate free and press wine fractions (*13*).

Enrichment of Diacetyl Components

The desire for richer, less fruity and more complex red and white table wines has led to a growing body of research and applied techniques associated with increasing the diacetyl, or 'buttery', character. This is achieved by one major pathway - malo-lactic (ML) bacterial fermentation, generally initiated by culture inoculations in young wines. The reaction is a catabolic pathway in which L-malic acid is enzymatically oxidized to L-lactic acid, carbon dioxide gas, and energy. Total titratable acidity is significantly reduced and pH increased in the process. By-products from the conversion include acetoin and 2,3-butanediol, acetic acid, and the diacetyl component of principal interest.

Most cultured lactic acid bacteria are found in the species, *Leuconostoc oenos* and *Lactobacillus* spp., existing as facultative anaerobes. This may be borderline as some studies indicate that higher O_2 concentrations can inhibit ML fermentation - and other equally sound research indicates quite the opposite. While the correlation of free oxygen analysis can reveal historical patterns for ML behavior, there is no finite set of predictors. Such is the fastidious nature of these microorganisms. More definitive wine bacteria characteristics are provided in Table 1.

More consistent is the importance of lower SO_2 levels, moderate temperature, and higher pH range required for ML fermentation. Free SO_2 levels greater than 20 mg/L can be expected to slow or stop activity of the bacteria. A rather narrow window of temperature, 20-25°C is generally considered ideal. The popular PSU-1 culture tolerates comparatively lower pH levels - in the 3.20-3.40 range, while the equally accepted ML-34 strain is more adaptable in ranges higher than 3.40.

Cultured ML organisms are heterotrophic and therefore unable to synthesize important nutrients from most naturally-occurring sources. Consequently, young wines typically require additions of B complex vitamins - often in the form of yeast cytoplasm extracts refined from

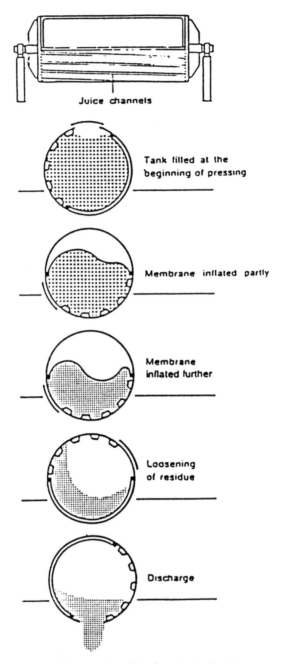

Figure 1. Operation of a Typical Tank Press
Adapted from: Food Technology International Europe (13)

autolysates. It is from this rationale that an increasing number of vintners leave young wines 'sur lies', or on fermentation sediment rich in dead yeast cells. Some winemakers actually inoculate wines with ML bacterial during primary yeast fermentation in order to take maximum advantage of higher yeast populations and therefore greater available autolysate substrates. Opinions differ in support of this as some evidence exists indicating active yeast growth may retard ML bacterial development.

Proper development of diacetyl and related complexity components require wine exposures open to the hazards of various other types of infection and malady. A very sound program of assurance is needed. Simple paper chromatograms can suffice in monitoring the malic-lactic conversion, but early indicators of acetic acid-ethyl acetate volatile acidity from *Acetobacter* spp. are essential. Diacetyl can be quantified by a distillation and spectrophotometric procedure provided by Zoecklein (*15*). Volatile acidity is adequately and easily measured by the traditional Cash distillation procedure (*11*).

Identification and Control of Hydrogen Sulfide and Mercaptans

The development of only trace quantities of hydrogen sulfide and mercaptans in wine results in foul odors which can seriously detract from wine quality and value. Elemental sulfur, often traced to vineyard spray residues, is the principal precursor. With continued reduction of labels authorized for viticultural use, reliance upon sulfur compounds in the vineyard has grown. This increased usage has given rise to a commensurate advance in the magnitude of problems associated with elemental sulfur.

Winemakers encourage maximum time intervals, typically not less than a month, between sulfur treatments in the field and grape harvest. Applications of colloidal sulfur generally result in heavier elemental sulfur residues, although the dusting, precipitated, and wettable forms are major sources, too. Progressive winegrowers are now applying micronized sulfur (particles which are less than 10 μm) dissolved in water. Another source of elemental sulfur is the residue from burning sulfur sticks inside wooden aging and storage vessels. Alternative methods of disinfecting, such as the use of live steam or potassium meta-bisulfite for shorter term, and the burning of dripless sulfur devices for longer term, are now widely employed.

According to Eschenbruch (*4*), the normal growth pathway of wine yeasts requires about 5 mg/L of sulfate for reduction to elemental sulfur in cell metabolism - from grape musts that contain up to 700 mg/L of available sulfate. The synthesis of hydrogen sulfide from this reduction is, thus, inherent with fermentation. Figure 2 portrays the formation of H_2S and mercaptans by yeasts.

Table 1. Physiological Characteristics of Wine Bacteria

Organism	Gram reaction	Catalase	Oxygen reqs.	Major endprd.	Sporulation
Gluconobacter	neg.	+	Aer.	Acetic	neg.
Acetobacter	neg.	+/−	Aer.	Acetic	neg.
Lactobacillus	pos.	−	Aer./ana.	Lactic	neg.
Leuconostoc	pos.	−	Fac./ana.	Lactic	neg.
Pediococcus	pos.	−	Aer./ana.	Lactic	neg.
Bacillus	pos.	+	Aer.	Several	pos.

Adapted from: Production Wine Analysis 1990 (15)

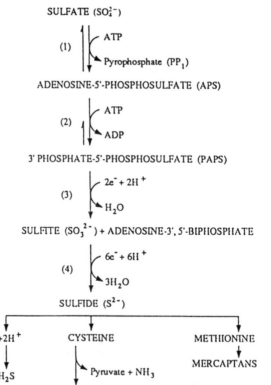

Figure 2.   Formation of Hydrogen Sulfide and Mercaptans by Yeasts
Adapted from: Production Wine Analysis 1990 (15)

Optimal removal of suspended grape skin fragments, pulp and other solids in white juice prior to fermentation commensurately reduces one of the highest sources of elemental sulfur. Wild yeasts and tumultuous fermenting yeasts are associated with greater levels of detectable H_2S 'rotten egg' odor. Epernay 2 and Prise de Mousse yeasts synthesize lesser degrees of H_2S, while the Montrachet and Steinberg strains generally result in comparatively higher levels. Deficiencies in essential vitamins, as well as FAN, are associated with increased production of H_2S due to the stimulation of proteolytic deamination reactions triggered by stressed yeasts.

Mercaptans also result from elemental sulfur and, even at trace levels, are significantly more pungent than H_2S. The typical sensory response to mercaptans is a 'skunky' odor, although it is the methyl and ethyl mercaptan forms that are found in wine, while skunk spray is comprised of the n-butyl mercaptan form. Methyl mercaptan, the most common wine malady, is formed from the amino acid methionine - typically as result of yeast deamination due to free ammonium nitrogen deficiency stress. Ethyl mercaptan is formed by the presence of acetaldehyde catalyzed by H_2S to form the intermediate thioacetaldehyde and water (15).

Contemporary winemakers use the assurance of adequate essential vitamins and FAN availability in fermenting musts as the most effective safeguard against the formation of H_2S and mercaptans. While the analysis of these constituents provide a difficult hurdle for most commercial enologists, reasonably good correlations can be made by using ion selective electrodes with expanded-scale pH/mV meters for the determination of ammonia.

Early on detection is another key part of control. Zoecklein et al. (15) have devised a method of separating H_2S and mercaptans with copper sulfate and cadmium sulfate reagents in the laboratory to facilitate better sensory determination more quickly and easily. Cooler fermentation temperatures are conducive to lower H_2S formation and can allow for better sensory detection early on in fermentation - offering the possibility for aeration or CO_2 to help sparge H_2S. Aeration during racking can also help to volatilize H_2S, but at the risk of oxidation and browning - less a concern for reds than whites.

Bentonite fining can reduce H_2S, but does not generally effect a complete removal. Sulfur dioxide additions can also reduce H_2S by oxidizing it back to elemental sulfur. The most effective and commonly used treatments are with appropriate prescriptions of copper sulfate - which can reduce or remove both H_2S and mercaptans.

Better Applications of Finings

Despite the improved mechanical virtues of crusher-stemmers and presses, some red grape varieties under

unusual growing conditions can still be problematical in
releasing phenolic compounds and suspended solids. One of
the most persistent problems in white wine fining is the
determination of soluble protein content. Far greater
levels of extracted phenols eliminates a concern for
protein instability in red wines.

The use of finings, or clarification agents, goes
back several centuries - although much of the technology
in understanding their use is new. One traditional
problem has been the excessive use of finings,
particularly bentonite, casein, and gelatin, all of which
can severely reduce color and flavor. On the other hand,
deficiencies in prescribing finings can result in
suspended solids remaining which can be substrates for
eventual enzymatic degradation of wine. These can result
in an unpleasant 'mousey' aftertaste yet remaining for
research to precisely identify. Underfining can also
allow copper and iron to remain in ionic form and cause
later haziness or 'casse' (12).

Brenna and DeVecchi (2) have reported the development
of a new assay for the determination of soluble proteins
in order to verify effectiveness of fining in white wines.
Wine samples are gel filtered in mini-columns packed with
Sephadex[R] in a tartrate-ethanol solution. Phenols are
bound by a dye on the column and thusly eliminated from
interference. The analysis is simplistic and low cost,
taking about one half hour to perform - although multiple
columns can be easily employed to render a series of
assays individually more time efficient.

Some vintners have returned to a more traditional
approach employed in Europe - that of 2 to 4 egg whites
beaten to a froth and then added to each 100 gallons or so
of wine in barrel. Adding a light dose of tannic acid
beforehand helps to prevent the formation of degradation
compounds from phenolics in the wine. The egg white
reaction with the tannic acid forms a fine granular
suspension that attracts suspended particles that become
a mass heavy enough to precipitate. While this is a
rather fascinating operation to witness in romantic old
wine cellars, it is at best difficult to manipulate and
time intensive (12).

One of the newest fining agents is colloidal silica -
perhaps better known as 'kieselsol'. Use of this compound
requires a prior treatment of protein fining, such as
casein, gelatin, or egg white mentioned above, but at a
reduced rate. Various grades and types of kieselsols have
widely replaced bentonite in Europe and increasingly so in
the U.S. It can be added directly to the wine without the
time-consuming slurry preparations required by bentonite
and quickly coagulates to embrace any phenol degradation
products that may have resulted from the protein fining
addition beforehand. The important advantages of
kieselsol are its reluctance to reduce color and flavor
unless they are already precariously unstable - and a
brilliancy that has encouraged an increasing amount of
bottling without filtration (12).

Quality Improvements by Membrane Separation Techniques

Contemporary psychographic tastes and preferences have resulted in an increasing concern about the human diet and the amount of alcohol consumed in wine drinking. Consumer demand for low and no alcohol wines grows in appeal to commercial wine suppliers and has generated an expanded interest in membrane separation technology. As this body of knowledge has grown, other uses of membrane separation have been employed for improving wine quality.

The principle involves the separation of one liquid into two of varying properties by virtue of a semipermeable barrier which controls the velocity of various molecules between its two sides. In short, its a molecular sieve. Depending upon the specific properties of this sieve, membrane separation is classified thus: 'Microfiltration' (MF), 'Ultrafiltration' (UF), and 'Reverse Osmosis' (RO). Retention ranges are provided in Table 2.

Membrane separation units typically consist of a holding vessel from which untreated liquid is stored; a pump that feeds the liquid at a proper rate and velocity; and a module that houses the membranes. This type of system is illustrated in Figure 3.

Juice or wine is appropriately pressured upon the inlet to the module and a <u>permeate</u> stream free of solids retained by the membrane, passes through and is collected. The <u>retentate</u> is the concentrated liquid retained which flows over the membrane and is collected separately. The rate of flow is called the <u>flux</u> and is directly proportional to lower pump pressures (permeate flux at higher pressures is non-pressure dependent) and inversely proportional to membrane resistance (*13*).

MF discriminates by particle size and has been successful in eliminating the need for bentonite fining, as well as centrifugation and filtration. This also provides the added advantages of conserving delicate color and flavor components, while also being more cost effect than more conventional processes. Membranes gauged for lower molecular weight solutes can also bar the passage of microorganisms - resulting in sterile wine products and lesser dosages of traditional wine preservatives (*14*).

UF separates on the basis of chemical structures in solution with ranges typically designed for large molecular weights such as colloids and polymers - finding excellent applications in reducing protein instability and separation of color pigments that have browned due to oxidation (*14*). Flores et al. (*5*) studied the effects of UF on aroma and flavor characteristics of Johannisberg Riesling and Gewurztraminer wines. Laboratory and plant trials generally failed to show a significant loss of fruit aroma by extensive sensory evaluations.

RO membranes inhibit the flow of comparatively lower

Table 2. Membrane Retention of Wine Components

Membrane Separation Process	Relative Retention of Wine Components		
	None	Partial	Total
RO	Water, ethanol, acetic acid, etc.	Acetaldehyde, glycerol, simple phenolic compounds, aromatic components, esters, etc.	Salts, glucose, flavonoids, macromolecules, microorganisms, and suspended particulates.
UF	All the soluble components with MW below 1000: amino acids, simple phenolic compounds, flavonoid monomers, etc.	Oligomers of flavonoids, peptides and all components with MW between 1000 and 10000.	Compounds with MW greater than 10000: proteins, polysaccharides, polymers of falvonoids, microorganisms and suspended particulates.
MF	All compounds with MW below 100000. All the soluble components.	Colloids with MW between 100000 and 10000000.	Colloids of MW greater than 1000000, microorganisms and suspended particles.

Adapted from: Proceedings of 7th International Conference in Food Science & Technology 1992 (14)

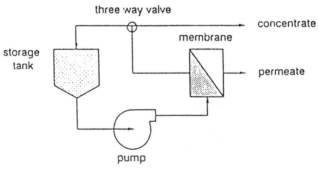

Figure 3. Membrane Separation System
Adapted from: Proceedings of 7th International Conference in Food Science & Technology 1992 (14)

molecular weight solutes and serves to concentrate juice or wine to products that are heavier-bodied and richer in color and flavor. The RO application also has the advantages of cool temperature operation as opposed to evaporator concentration and far less demand for energy. This can also supersaturate ionic potassium and tartaric acid in solution - aiding in faster precipitation of unstable potassium bitartrate crystals. Deposits of these argols frequently form on equipment surfaces causing restricted product flow. Electrodialysis, consisting of electrolytes migrating through an anion impermeable membrane on one side and a cation impermeable membrane on the other, both under the effect of an electric field, can minimize the presence of potassium ions. Electrodialysis can also be effective in the reduction of acetaldehyde and sulfurous acid content in wine (14).

For alcohol reduction or removal, the system is designed with a membrane selective only for alcohol permeate, while the remaining constituents form the retentate. Alcohol concentration levels are controlled by introducing a flow of water on the permeate side of the membrane - carrying away alcohol molecules and thus reducing the alcohol concentration on the retentate (product) side (14).

Fully ripened grapes cultivated in cooler climates are often deficient in desired sugar content. Sensitivity to this can be exemplified by the German wine classification system which is based upon sugar/ripeness parameters - as compared to a geographical basis in most other European countries. The practice of sugar additions, or Chaptalization, has long been criticized as a breach of ethics - and commercially prohibited in California. A recent study conducted Duitschaever et al. (3) concludes that Johannisberg Riesling wines made by musts concentrated by RO were higher in quality than those which were Chaptalized. Premature grapes suffering from an abbreviated growing season are, however, often excessively high in acid and phenol constituents - a situation aggravated even further by concentration. Pompei (7) suggests that these excesses can be relieved by rectification prior to RO treatment, but the lack of flavor development in green fruit often fails to justify the cost of such processing.

Better Economies in Bottling and Packaging

With the increased investment in machinery and maintenance expense necessary to "in-house" bottle and package wine many vintners have opted to contract mobile services to perform this function (13).

Apart from the cost aspects mentioned above, mobile bottling also permits better efficiencies of both time and space. This tends to increase in importance inversely with winery size. The smaller the operation, the less time is usually devoted to bottling and packaging -

increasing cost on a per unit output basis. It follows
that such equipment is thus oversized - influenced
primarily by limited selections in down-sized model
availabilities from manufacturers. The area replaced by
mobile bottling in most small wineries ranges from 500 to
more than 1,000 square feet. In some of the more
expensive vinicultural real estate locales this can exceed
$100,000 of investment, with a similar outlay for the
actual bottling and packaging equipment. Opportunity cost
for a modest 10,000-case winery can easily be double the
cost of mobile bottling. Setup, trial and error, waste,
as well as training and retraining, are also major cost
jeopardies inherent with small winery in-house bottling
and packaging operations.

Custom bottling firms quote prices for bottling and
packaging services on site at less than $2 per case -
significantly less if only a front label is applied. In
that the mobile bottling equipment is regularly used, it
is effectively maintained and, equally important, replaced
with more productive and efficient units. The vintner
thus benefits from state-of-the-art equipment and material
employed to optimize wine quality. Quantity discounts are
involved with pricing, as are mileage and filter usage
considerations. Experienced professional operators and
guaranteed results are additional attractive appeals in
favor of the custom service.

Scheduling can be problematical with mobile bottling,
although competition and advance planning continue to
reduce this disadvantage. Conscientious vintner
preparation prior to mobile arrival is critical as winery
caused down time may be charged at rates of $150 per hour
and more.

Unfortunately, this type of service remains limited
to only the more concentrated winegrowing regions of the
U.S. The Bureau of Alcohol, Tobacco, and Firearms,
enforce production location regulations which greatly
inhibit inter-vintner cooperation in sharing equipment.

Summary

Time-honored methods and techniques continue to be
improved by the compelling influence of problems explained
and solved by science. The application of current
technology across every discipline and phase of wine
production results in contemporary wines being the highest
quality ever released by the industry. Technological
developments in progress promise to provide new
applications for even greater wine appreciation.

Literature Cited

(1)   Boone, C.; Bussey, H.; Greene, D.; Thomas, D.Y.;
       Vernet, T. Yeast killer toxin: site directed
       mutations implicate the precursor protein as the
       immunity component. *Cell* **1986**, *46*, 105-13.

(2)   Brenna, O.; DeVecchi, S. Evaluation of protein and phenolic content in must and wine. 1. Assay of soluble proteins. *Ital. J. Food Sci.* **1990**, *4*, 269–272.

(3)   Duitschaever, C.L.; Alba, J.; Buteau, C.; Allen, B. Riesling wines made from must concentrated by reverse osmosis. I. Experimental conditions and composition of musts and wines. *Am. J. Enol. Vitic.* **1991**, *42*, 19–25.

(4)   Eschenbruch, R. H_2S formation – The continuing problem during winemaking fermentation technology. *Australian Society of Viticulture and Oenology Proceedings* **1983**, 79–87.

(5)   Flores, J.H.; Heatherbell, D.A.; Henderson, L.A.; McDaniel, M.R.   Ultrafiltration of wine: Effect of ultrafiltration on the aroma and flavor characteristics of white riesling and gewurztraminer wines. *Am. J. Enol. Vitic.* **1991**, *42*, 91–96.

(6)   Panagiotakopoulou, V.; Morris, J.R.   Chemical additives to reduce browning in white wines. *Am J. Enol. Vitic.* **1991**, *42*, 255–260.

(7)   Pompei, C.   Le sucre de raisin.   Aspects technologiques. *O.I.V. Bulletin* **1982**, *611*, 25–52.

(8)   Robredo, L.M.; Junquera, B.; Gonzales-Sanjose, M.L.; Rarron, L.J.R.   Biochemical events during ripening of grape berries. *Ital. J. Food Sci.* **1991**, *3*, 173–80.

(9)   Sims, C.A.; Bates, R.P.; Mortensen, J.A.   Effects of must polyphenoloxidase activity and timing of sulfite addition on the color and quality of <u>Vitis rotundifolia</u> and <u>Euvitis</u> hybrid white wines. *Am. J. Enol. Vitic.* **1991**, *42*, 128–132.

(10)   van Vuuren, H.J.H.; Jacobs, C.J.   Killer yeasts in the wine industry: A review. *Am. J. Enol. Vitic.* **1992**, *43*, 119–128.

(11)   Vine, R.P.   In *Commercial Winemaking*; AVI/Van Nostrand Reinhold: New York, NY, 1981.

(12)   Vine, R.P.   Making sense of fining. Indiana Wine Grape Council Wine Lines, Summer 1992.

(13)   Vine, R.P.   The use of new technology in commercial winemaking. *Food Technology International Europe* **1987**, 146–149.

(14)   Vradis, I.; Floros, J. Membrane separation processes for wine dealcoholization and quality improvement. Proceedings of 7th International Conference in Food Science & Technology, 1992.

(15)   Zoecklein, B.W.; Fugelsang, K.C.; Gump, B.H.; Nury, F.S. In *Production Wine Analysis*; AVI/Van Nostrand Reinhold: New York, NY, 1990.

RECEIVED May 6, 1993

Chapter 9

Biotechnological Advances in Brewing

Marilyn S. Abbott, Tom A. Pugh, and Alastair T. Pringle

Corporate Research and Development, Anheuser–Busch Companies, Inc., 1 Busch Place, St. Louis, MO 63118

A variety of biotechnological tools have been applied to improve the ingredients of the brewing process. Using these tools agricultural materials have been developed that are free of viruses, have improved agronomic yields, or are resistant to disease. Brewer's yeasts have been constructed with novel properties such as the ability to ferment normally unfermentable carbohydrates, chill-proof beer, or degrade beta-glucans. Yeasts have also been developed that produce less diacetyl, have altered flocculation properties, or are resistant to contamination. Although there are many advantages to biotechnologically improved agricultural materials and yeast, these advantages must be weighed against regulatory, legal and consumer concerns.

The word biotechnology made its debut in 1919 when it appeared in the book by Karl Ereky entitled *Biotechnology of Meat, Fat, and Milk Production in a Large Scale Industrial Enterprise*. Ereky broadly defined biotechnology as all processes that create products from raw materials using living organisms. Using this broad definition, it can be said that brewing is one of the oldest biotechnologies, since beer was being brewed by the Summarians as long ago as 6000 BC.

Brewing changed little over the next 7000 years until scientists became interested in the brewing process. In 1680, the Dutch microscopist, Anton Van Leewenhoek, was the first to observe brewer's yeast. In the early part of the 19th century Cagniard de la Tour of France, and Schwann and Kutzing, both from Germany, proposed that the products of alcoholic fermentation (ethanol and carbon dioxide) were made by a microscopic form of life. It was not until the 1860's, however, that Louis Pasteur was able to demonstrate that yeast was responsible for the fermentation of alcoholic beverages. Pasteur was also instrumental in saving the French beer industry by identifying bacteria as the agents of beer spoilage and then developing pasteurization to preserve bottled beer.

0097–6156/93/0536–0150$08.75/0
© 1993 American Chemical Society

The next significant advance was a technique for isolating pure cultures of yeasts, developed in 1883 by Kristian Emil Hansen of the Carlsberg Laboratories, Denmark. Prior to Hansen's innovation, brewers used undefined mixed cultures of yeasts with somewhat unpredictable results. Since that time numerous advances have been made in biotechnology that have benefitted the brewing industry. In the rest of this chapter we summarize the advances that have affected the ingredients of beer: yeast, barley, corn, rice, and hops.

Application of Biotechnology to Agricultural Materials

In the following sections biotechnologies that can be applied to agricultural raw materials for brewing will be discussed. In brewing beer, barley, rice, corn and/or other cereal grains are sources of carbohydrates for the yeast, which converts these to ethanol and carbon dioxide. Hop resins and oils are extracted from the hop flowers during wort boiling and impart bitterness and aroma, respectively. These agricultural raw materials represent a substantial cost to the brewer, thus a great deal of effort has been expended to improve agronomic yields, brewing efficiencies and flavor characteristics. Here, some new methods of improving brewing agricultural materials will be discussed together with specific applications.

Plant Breeding. Figure 1 depicts the life cycle of a typical flowering plant. The mature plant produces flowers bearing the male and female reproductive structures, the stamen and pistil, respectively. The stamen consists of a filament supporting the anther, which contains the pollen. Mature pollen is released from the anthers and is deposited on the upper portion of the pistil where it germinates producing a pollen tube. The pollen tube then grows through the tissue of the pistil to the base where the ovule is located. Male and female nuclei, which each contain half the normal number of chromosomes (haploid), fuse producing an embryo that contains a full complement of chromosomes (diploid). The embryo matures within the seed, and upon germination, grows into a new plant.

In the process of hybridization, plant breeders control fertilization by transferring pollen from one plant to the pistil of another to produce seed progeny containing mixtures of traits from both parents. The breeder then selects those progeny plants that exhibit the desired combination of traits. These basic techniques of hybridization and selection have been used by plant breeders for nearly one hundred years. Modern breeding programs have improved the efficiency of these procedures by modifying selection schemes and by using greenhouses and off-season locations to speed the development of new varieties (1). These approaches have been so successful that, for example, new varieties have increased barley yields in the Midwest by nearly 100% over the past 40 to 50 years (2).

Micropropagation. Commercial cultivars of some crop species, such as potato, grape, and hop do not have two identical copies of each gene. Seed

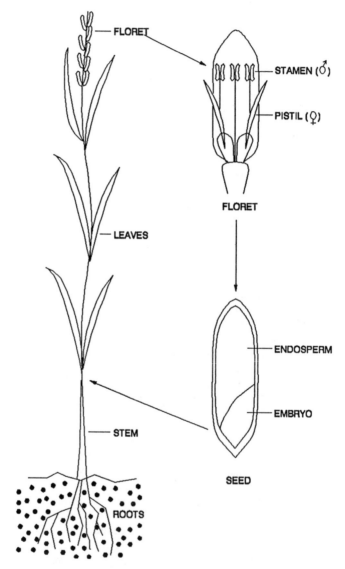

Figure 1.   Life cycle of a typical flowering plant.

progeny of these plants, like those of hybrid corn, are not uniform and for many traits do not resemble the parent variety.   These crops are propagated vegetatively through subdivision and multiplication of portions of the parent plant to maintain genetic uniformity. Micropropagation is a method for rapid production of unlimited numbers of genetically identical copies of an individual plant. This is accomplished by excising the undeveloped meristems (growing points) of a plant and culturing them *in vitro* on a medium that induces the formation of multiple new shoots. The meristems from each of these new shoots are excised and the process is repeated until the desired number of shoots is produced. The shoots are then transferred to a root-inducing medium, after which plants can be transferred to soil and then to the field.   Using micropropagation techniques several hundred thousand plantlets can be produced from one small hop plant *(3)*. Many new commercial hop varieties are now being introduced through micropropagated plant material.

Virus Testing.   Viral diseases can significantly decrease crop yield and quality in both barley and hop. Some of these diseases are seed-borne in grains or are transmitted through vegetative propagation of crops such as hop. Since there are few effective chemical treatments for viral diseases, testing planting material is an important precaution. Rapid virus testing is accomplished using an antibody-based method such as the enzyme-linked immunosorbent assay (ELISA). An ELISA utilizes antibodies that bind to the virus coat protein found in extracts of infected plant tissue. Sensitive ELISA tests have been developed for many different plant viruses that infect barley or hop.

In several states, barley seed is tested by ELISA for Barley Stripe Mosaic virus before it can be certified for planting. ELISAs are also used for routine screening of hop planting stock for viruses such as Prunus Necrotic Ringspot, Hop Latent, and Hop Mosaic. Virus-free hop plantlets can be produced from infected material by heat treating of shoots, to inhibit virus multiplication, followed by excision and *in vitro* culture of the terminal bud meristems *(3)*.

Viroids are small pieces of RNA that can infect plant cells. Two viroids, Hop Stunt *(4)* and Hop Latent *(5)*, have been shown to infect commercial hop cultivars. Since viroids do not have a protein coat, they cannot be detected using ELISA; however, the viroid nucleic acid (RNA) can be detected using a technique called "dot blot hybridization". The Horticultural Research Institute (UK) has recently begun using this procedure to screen hop plantings in England *(6)*.

Varietal Purity by DNA Fingerprinting.   After a plant variety has been selected, tested, and multiplied for commercialization, a breeder must decide whether to protect his property rights over the new variety and how best to accomplish this. Plant varieties can be protected by patenting or by certification through the Plant Variety Protection Act (PVPA). The type of protection provided by each approach is somewhat different; however, both require evidence that a new cultivar is clearly "distinct" from existing varieties *(7)*. To obtain this evidence, many plant breeders are turning to "DNA fingerprinting" technologies, which can readily distinguish varietal differences in corn and rice

(8, 9). In fact, the technology has gained such legitimacy, in plant as well as human criminal cases, that DNA fingerprinting evidence was recently used in an ownership dispute over a celery variety *(10).*

Selection Of Improved Plant Varieties

Anther Culture. The most immediate applications of biotechnology to plant breeding are in the selection of improved progeny. To develop new commercial cultivars, plant breeders must be able to identify and select, out of a heterogeneous population, those few individuals that express desired traits. This can be particularly difficult with traits that are only partially expressed in the heterozygous condition (only one copy of the desired gene). Traditional breeding practices assume that approximately ten rounds of self-pollination are required to reach nearly complete homozygosity (both copies of each gene are identical).

This time-consuming process can be circumvented by anther culture, which produces "doubled haploid" plants having two identical copies of each chromosome (Figure 2; for simplicity, only two types chromosomes are depicted). Anthers, the pollen compartments of the stamens, are removed from the florets and placed on a medium that induces each immature pollen grain to develop into a mass of cells termed a callus. The calli are transferred to another medium where they develop into green plants. Since pollen grains are haploid (contain only one copy of each chromosome), plants from anther culture should also be haploid; however, the formation of an extra set of chromosomes often occurs spontaneously, or it can be induced by a chemical treatment. This results in a normal plant that has two identical sets of each chromosome (homozygous diploid).

Using anther culture, the breeder can evaluate homozygous lines five to six generations earlier than would be possible using traditional procedures. In addition, these plants can be selected for further crossing or testing with the knowledge that they will show no further change in the expression of any traits in subsequent generations. Anther culture is now used extensively in rape seed breeding programs *(11)* and is beginning to be utilized for rice and barley variety development *(12).*

Marker-Assisted Selection. Many of the brewing traits are very difficult to select in early generations because they are process-related traits such as amount of extract or type of flavor. This complexity forces the breeder to carry lines for several generations until semi-commercial-scale analyses can be performed. Marker-Assisted Selection (MAS) is a technique that addresses this problem. The procedure depends upon close genetic linkage (proximity on a chromosome) between a desired trait that is difficult to screen for and an easily screened trait that can be used as a marker. If the marker gene is close enough on the chromosome to the desired trait, selection for the marker in each generation will increase the occurrence of the desired trait, even though it has not been directly selected. Even though MAS is a well-established concept for plant breeders, applications to date have been limited because few morphological or biochemical

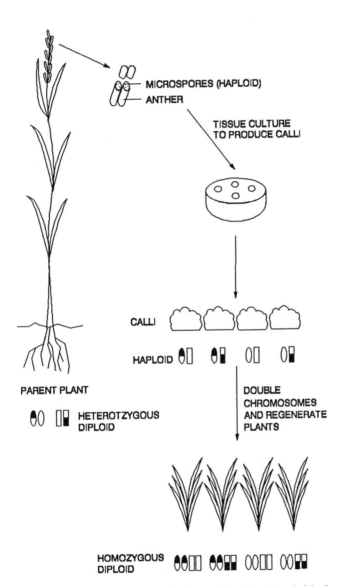

Figure 2.   Anther culture production of doubled haploid plants.

markers are available in most crops and often they are linked to undesirable traits.

Biotechnology has renewed interest in the MAS approach through the development of Restriction Fragment Length Polymorphism (RFLP) markers. RFLPs correspond to small (sometimes single base-pair) differences in the DNA of different cultivars. RFLPs are identified by isolating DNA and digesting it with enzymes specific for particular base sequences (depicted as scissors, Figure 3). Samples of DNA from different individuals will produce different sized fragments that are detected as RFLPs.

RFLPs are a great improvement over morphological or isozyme markers because hundreds of them can be identified and located on the chromosomes (mapped) permitting assignment of markers with very close linkage to almost any trait of interest *(13)*. Linkage maps of RFLP markers have been constructed for many crop species and are being used to select varieties with improvements in processing qualities, such as soluble solids in tomato *(14)*; and agronomic traits, such as grain yield in corn *(15)*.

Further simplification of MAS screening can be provided by the Randomly Amplified Polymorphic DNA (RAPD) marker procedure developed by Williams *et al. (16)*. This procedure reduces the number of steps and eliminates the need for the radioisotopes usually used in RFLP analyses. Martin *et al.* recently demonstrated the ease with which markers linked to a trait of interest can be detected using RAPDs to identify genes for resistance to a bacterial pathogen in tomato *(17)*.

MAS procedures are being developed for both hop and barley. Researchers at Wye College (UK) have begun to identify RAPD markers for important traits in hop with the intention of using them to select for varieties with improved disease resistance and brewing qualities *(6)*. For barley, a group of American and Canadian researchers are collaborating on the North American Barley Genome Mapping Project. The goal of this project is to produce a genetic map of barley comprised of RFLP and RAPD markers that will be available to all barley breeders. At the time of writing, this group has produced a map containing 296 markers *(18)* and has begun to correlate the markers with important agronomic, malting, and brewing traits. The development of DNA-based screening techniques will make it possible for hop and barley breeders to screen individual seedlings for complex traits such as insect and disease resistance, flavor, and brewing extract.

Mutation Breeding. Plant breeders use mutation methods to increase the amount of variation within the germplasm in their programs. The most commonly used mutagens for plants are X-rays, ethyl methanesulfonate (EMS), and sodium azide, which are applied as seed treatments *(19)*. The treated seeds are then planted and the resultant plants are evaluated for traits of interest. Barley mutants have been produced by EMS treatment that have very low levels of polyphenolic compounds called anthocyanogens. Anthocyanogens are largely responsible for the formation of chill-haze in beer, and it has been found that beers brewed with anthocyanogen-free malt show significantly improved chill-haze stability *(20)*.

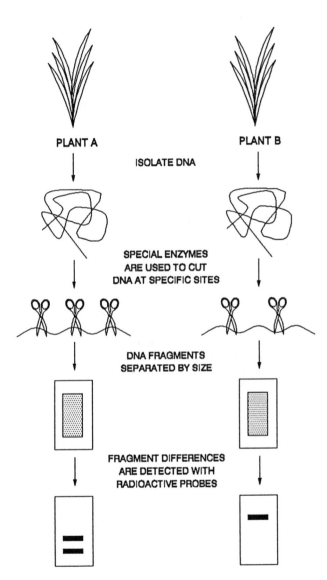

Figure 3.   Detection of a restriction fragment length polymorphism (RFLP).

A more unusual type of mutagenesis is somaclonal variation, which describes the enhanced genetic variability that can arise during tissue culturing *(21)*. Tissue culture is a procedure in which a portion of a plant is removed and induced to grow on a defined medium in the laboratory. Depending on the culture conditions, the cultured plant material can produce either callus or a fine suspension of cells called a suspension culture. Further changes of culture medium then cause differentiation into shoots and/or roots. *In vitro* culture of plant tissues sometimes leads to chromosome instability resulting in heritable duplications, translocations, point mutations, etc. Somaclonal variation has been used successfully with many different crop species to produce variant plants, e.g., short stature, resistance to disease (22) or to insects *(23)*, or tolerance of acid soils *(24)*. In some cases, a purified toxin from a pathogen can be included in the culture medium, thus providing an early selection method for disease-resistant lines. For example, corn lines resistant to leaf blight disease have been selected in this way *(25)*.

An important prerequisite of somaclonal variation is a procedure for regenerating plants of the desired species from tissue culture. Fortunately, these techniques are becoming available for brewing agricultural materials. Hop plants reportedly have been obtained from callus cultures, and a somaclonal variation program is now underway at Wye College (UK) *(6)*. In addition, Jahne and co-workers have recently described the regeneration of barley plants from cell suspension cultures *(26)*.

Current Status of Transformation Technologies. Often a breeder wishes to make only a single change in an otherwise acceptable cultivar. This is very difficult to achieve using traditional breeding methods, as most of the genes are recombined whenever a cross is made. Genetic engineering offers the potential to add or delete a single gene while maintaining the desirable traits of commercial varieties. During the past decade, procedures such as PEG precipitation, electroporation, *Agrobacterium* infection, and biolistics have been developed to introduce new genes into plants (transformation).

The first step for both PEG precipitation and electroporation is the removal of the plant cell walls to form protoplasts. The protoplasts then are mixed with DNA containing desired genes and are subjected to either an osmotic (PEG) or an electric (electroporation) shock, which transiently opens holes in the protoplast membrane large enough to permit the DNA to pass through. The transforming DNA is incorporated into the host chromosomal DNA through an unknown mechanism. The next step is to produce whole plants from the protoplasts. In some plants, such as rice, this can be done fairly readily *(27)*; however, in many others, such as barley, regeneration of plants from protoplasts has been unsuccessful despite substantial effort.

Agrobacterium is a genus of soil bacteria that infects several types of plants causing tumorous growths *(28)*. *Agrobacterium* induces this tumor formation by transforming the plant cells with genes that encode plant growth hormones. Researchers have manipulated certain *Agrobacterium* species to eliminate the tumor-causing genes and replace them with other genes that they wish to transfer into plants *(29)*. This system works very well for certain types of plants such as tobacco, tomato, and potato; however, most major crop species (i.e. rice, corn, wheat, and barley) are not susceptible to infection by *Agrobacterium*.

A biolistics approach has been widely used for transformation of crop plants. In this procedure, DNA is coated onto tiny metallic particles that are fired into the tissue to be transformed using a modified gun. Many of the cells surrounding the site of impact are destroyed, but some of the transformed cells survive and regenerate into plants. To date, transformed rice *(30)*, corn *(31)*, and wheat *(32)* plants have been produced using biolistics. Unlike the PEG and electroporation techniques, biolistic methods do not require the use of protoplasts, which should make this approach applicable to more species and to more cultivars within species *(33)*.

Unfortunately, stable transformation has not yet been reported for barley or hop. Hop may be susceptible to *Agrobacterium* infection, which could pave the way to relatively rapid development of a hop transformation system. Although a great deal of effort has been expended, no reproducible barley transformation system has yet been described. D'Halluin *et al.* *(34)* recently described a high-efficiency, cultivar-independent method for transforming corn in which immature embryos were briefly treated with enzymes to digest their cell walls and then electroporated in the presence of DNA. The embryos were subjected to a brief period of callus formation in tissue culture, after which transformed plants were regenerated from the calli. Barley researchers will no doubt evaluate this procedure in the very near future.

When practical transformation systems are developed for hop and barley, it will be possible to increase the expression level of existing traits such as hydrolytic enzymes or flavor compounds by inserting genes with enhanced control elements (promoters) *(35)*. Transformation also can be used to decrease the production of a naturally occurring gene product, e.g., a flavor compound or an indigestible protein. Using "anti-sense" technology, the plant is transformed with an inverted copy of the gene to be modified. Due to this inversion, a reversed or "anti-sense" copy of the gene message is produced in the transformed plant cells. Although the actual mechanism is not known, it is hypothesized that the anti-sense message binds to the normal message thereby preventing expression of the unwanted trait *(36)*.

Methods for Strain Improvement of Brewer's Yeast

While beer can be made from a number of agricultural materials, yeast is a key ingredient that cannot be substituted. The two main types of yeast are ale and lager, which ferment either at the top or the bottom of the fermentor, respectively. Several strategies have been used in attempts to improve brewer's yeast characteristics and have been reviewed in detail *(37-39)*. Although many techniques are available for yeast genetic manipulation, only those techniques that can be used to improve industrial strains of brewer's yeast will be discussed in the following section.

Genetic Organization. As in plants, the genetic information of the brewer's yeast *Saccharomyces cerevisiae* is stored and perpetuated as deoxyribonucleic acid

(DNA). The DNA is found in two locations in the cell (Figure 4): the nucleus and the cytoplasm *(40, 41)*. The nucleus contains genomic DNA, which makes up the majority of the total DNA content of yeast and is organized into chromosomes. Yeast contains at least one set of 16 chromosomes, maintenance of which is essential for cell viability. Brewer's yeasts typically have greater than three copies of each chromosome (polyploidy) and also have unequal numbers of each type of chromosome (aneuploidy) *(42)*.

The nucleus also may contain plasmids, which are small circular molecules of DNA. Many yeast strains contain 60 to 100 copies per cell of an indigenous plasmid called the 2μm circle *(43)*, which has been used as the basis of other plasmids developed for the genetic engineering of yeast using recombinant DNA technology (see Genetic Transformation).

The cytoplasm surrounds the nucleus and contains two additional genetic elements (Figure 4): the mitochondrion and the killer factor *(44, 45)*. There are approximately 35 mitochondria per cell and they enable yeast to grow on respirable carbon sources. Mitochondrial DNA is circular molecule and contains genes that encode products required for respiration and mitochondrial DNA replication. Mitochondria, however, are dispensable; strains that lack them are respiratory-deficient or petite, but they can grow on fermentable carbon sources.

Some yeast strains, called killer strains, contain the virus-like killer factor. Unlike other genetic elements, the genetic information of the killer factor is carried by ribonucleic acid (RNA) rather than DNA. The genes of killer factor direct the synthesis of a secreted protein toxin that is lethal to non-killer-factor-bearing strains. Most brewing strains, however, do not carry a killer factor.

Yeast Life Cycle. The life cycle of yeast is simple and consists of two phases: asexual and sexual (Figure 5) *(46-48)*. During the asexual phase, yeast grows and divides by budding. Yeast can grow indefinitely asexually; however, if growth conditions become poor and nitrogen is limited, yeast will sporulate and enter the sexual reproductive phase. The four haploid spores produced will germinate into haploid cells under favorable nutritional conditions. Yeast has two haploid cell mating types, which are classified *a* and α based on their mating preference. Only haploid cells of opposite types are capable of mating, so *a* and α cells may fuse to form a diploid cell. The resulting diploid cell can reproduce asexually by budding or, if starvation conditions are encountered again, it will sporulate and enter the sexual cycle.

Breeding or Hybridization. Four approaches have been used in attempts to breed yeast strains with improved characteristics: sexual hybridization, rare mating, cytoduction, and protoplast fusion.

Sexual Hybridization. In breeding programs using traditional sexual hybridization, a haploid donor strain with a desired trait conferred by a single gene is mated or crossed to a target strain. The hybrid diploid is then sporulated and the progeny are backcrossed to the parental strain. Backcrossing is repeated several times to ensure that only the desired trait is transferred from the donor strain to the target strain.

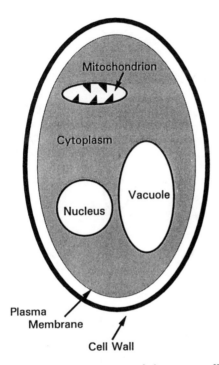

Figure 4.   Cytology of the yeast cell.

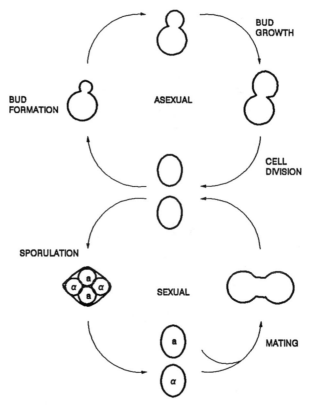

Figure 5.   Life cycle of brewer's yeast.

Application of sexual hybridization to brewer's yeast is difficult since they generally do not mate, sporulate poorly, and produce spores that have low viability. Despite these difficulties, sexual hybridization has been used successfully to modify brewing strains. For example, Gjermansen and Sigsgaard *(49)* optimized conditions to induce sporulation in a brewing strain. They found that sporulation occurred at a higher frequency at lower temperatures. They also enriched for spores by selectively killing vegetative cells with lytic enzymes. Although a majority of the spores were not viable, some mating-competent, spore-derived strains were isolated and used to construct a hybrid strain. Bilinski *et al. (50)* also used this technique to construct a lager and ale yeast hybrid strain that shared fermentation properties of both parents.

Rare Mating. On rare occasions, mating-competent cells arise spontaneously in brewer's yeast populations. Thus, brewing strains can be mated directly to haploid strains using high cell densities. Typically, this technique has been employed where a respiratory-deficient brewing strain is crossed to a respiratory-proficient haploid strain that requires a specific amino acid for growth (Figure 6) *(51, 52)*. Hybrids are selected by growing the mated mixture on medium without amino-acids but containing a respirable carbon source; only cells that mated are capable of growth. This method was used successfully by Tubb *et al. (53)* to incorporate the *DEX1* gene of *Saccharomyces diastaticus*, which encodes a secreted glucoamylase, into a brewing strain.

Cytoduction. Cytoplasmic elements, such as mitochondria and killer factor, can be transferred from a donor strain to a target strain using cytoduction (Figure 6). During normal mating, two haploid cells and their nuclei fuse to form a diploid cell *(47)*. If one of the haploids of the mating pair is a *kar1* mutant, however, mating is blocked at nuclear fusion *(54)*. When cell fusion occurs, the cytoplasms mingle but the nuclei remain separate. The nuclei of this transient pseudo-diploid or heterokaryon segregate into separate cells called cytoductants. These cytoductants contain the cytoplasms of both parents but the nucleus of only one.

On rare occasions, single chromosomes or plasmids can be transferred from one nucleus to the other in a heterokaryon *(55)*. In this way, the genetic constitution of individual chromosomes from brewing strains have been examined in genetically-defined haploid strains *(56)*. Alternatively, desired traits known to be associated with a particular chromosome can be transferred to brewing strains. In a recent example of this, Vezinhet *et al. (57)* used cytoduction to transfer the trait of increased flocculation from a donor strain to a wine yeast.

Protoplast Fusion. Protoplast fusion bypasses the need to obtain a yeast hybrid by sexual mating *(58, 59)*. In this technique, protoplasts are formed by enzymatically removing the cell walls of the two yeasts to be crossed. The protoplasts are then mixed in an osmotically-stabilized solution containing polyethylene glycol (PEG) and calcium to promote cell membrane fusion. The cell walls are then regenerated during growth on an osmotically-stabilized medium. Selective conditions used to recover fusion hybrids include

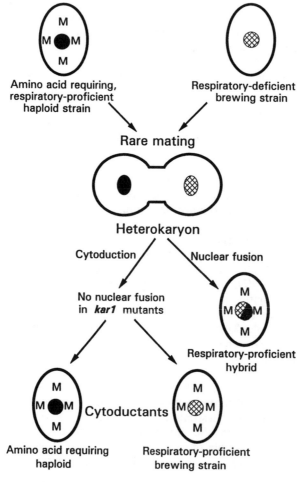

Figure 6.   Rare mating and cytoduction.

complementation of auxotrophic or respiratory-deficient mutations and resistance to metal ions or antibiotics *(60, 61)*. A drawback of protoplast fusion is that individual protoplasts can fuse with more than one other protoplast, sometimes producing unstable hybrids. Thus, caution must be used when selecting for, and assessing the genetic constitution of, fusion-mediated hybrids.

Selection. Genetic selection is the process by which individual cells with altered characteristics can be recovered from a cell population. Variant individuals arise in a population through mutations in the gene structure that may reduce, increase, or even abolish gene activity. Mutations occur rarely through spontaneous errors in DNA replication and repair, or they can be induced at significantly higher rates by mutagenic agents such as ultraviolet light (UV), ethylmethane sulfonate (EMS), and N-methyl-N-nitrosoguanidine (NTG). Most mutations are recessive and are not revealed in the presence of a normal copy of the gene; however, some mutations are dominant and are expressed regardless of the genetic background.

Since brewer's yeast is polyploid, it is difficult to select recessive mutations. Despite these difficulties, mutants of brewing strains with improved fermentation characteristics have been isolated *(62)*. Indicator and selective media have provided a useful means of identifying and isolating variants or mutants.

An indicator medium contains an ingredient that makes it easy to recognize a colony with an altered characteristic. For example, the selective medium ZYCM has been used to detect yeasts that produce little or no H_2S since these colonies are white compared to brown for the normal yeast *(62, 63)*. Similarly, WLN medium *(64)* has been used to select yeasts with increased metabolic activity that produce less diacetyl *(65)*.

Selective media contain ingredients that typically allow the growth of mutants and inhibit the growth of all other cells. Kielland-Brandt *et al.* *(66)* selected for dominant mutants that could grow on a medium containing the herbicide sulfometuron methyl (SM). SM-resistant mutants have an altered acetohydroxyacid synthase *(67)* that does not produce as much α-acetolactate and therefore produces less diacetyl.

Genetic Transformation. Genetic transformation of yeast has become a routine laboratory procedure *(68)*. Using this technique, DNA from other sources can be introduced into yeast. Target genes are most commonly introduced into cells on plasmids, which are convenient vehicles for the manipulation, amplification, and expression of cloned genes.

Plasmids. Most plasmids used in yeast transformations are capable of replicating in both bacteria and yeast. Propagation in bacteria is essential for large-scale purification of plasmid DNA as well as for routine manipulations of cloned genes. There are four basic types of plasmids that are used for yeast transformation (Figure 7): integrating, episomal, replicating, and centromeric *(68)*. Each of these plasmids contains a bacterial origin of replication (ORI), a bacterial selectable marker (for example Ap, which confers ampicillin resistance), and a yeast selectable marker (for example *URA3*, which satisfies the nutritional requirement for uracil in *ura3⁻* mutants). These plasmids differ in their ability to replicate in yeast, the basis of autonomous replication, copy number, and stability. Although other types of plasmids have been developed, they have not been used for transformation of brewer's yeast.

Yeast integrating plasmids (YIp) cannot replicate autonomously in yeast, but rather must integrate into chromosomes to transform yeast and therefore are stably transmitted to progeny. Yeast episomal (YEp) and replicating (YRp) plasmids replicate autonomously in yeast by virtue of 2μ DNA (from the 2μm circle plasmid) and an autonomous replication sequence (ARS), respectively. These plasmids are maintained at high copy number, but are unstable and are lost rapidly if transformants are grown under non-selective conditions. Yeast centromeric plasmids (YCp) also replicate autonomously, but are maintained at 1 to 2 copies per cell due to the presence of a centromere (CEN). YCp plasmids are much more stable than YEp and YRp plasmids, but can still be lost under non-selective conditions.

Yeast integrating plasmids are most useful in brewer's yeast strain development, since they are the most stable. These plasmids, however, contain bacterial DNA sequences. Current government regulations concerning the use of recombinant DNA technology in the food industry forbids the use of an organism into which DNA from a potential pathogen has been incorporated (69). Therefore, genetically engineered brewing strains must be constructed such that only the target gene is integrated.

Selectable Markers. After transformation has occurred, it is essential to be able to select transformed from non-transformed cells. Since brewing strains are polyploid and recessive selectable markers are not obviously expressed, plasmids used for transformation of brewer's yeast must contain dominant selectable markers. Several dominant selectable markers have been developed including resistance to copper ions, conferred by the yeast *CUP1* gene (70); resistance to the herbicide SM, conferred by the *SMR1-410* allele of the yeast *ILV2* gene (71); and resistance to antibiotics such as G418 (72) and hygromycin B (73), conferred by genes of bacterial origin.

Methods of Transformation. There are three commonly used methods for yeast transformation: spheroplast transformation, alkaline cation treatment, and electroporation.

In the spheroplast transformation procedure (74, 75), the yeast cell wall is partially removed by enzymatic digestion in an osmotically-stabilized solution. The spheroplasts are then exposed to plasmid DNA in the presence of PEG and calcium, which promotes passage of the exogenous DNA into the spheroplast. Spheroplasts are then plated on a selective medium. Despite being complicated and time-consuming, this procedure yields relatively high transformation efficiencies.

The second method is the treatment of intact cells with alkaline cations (76). Yeast cells are treated first with a lithium acetate or lithium chloride solution, then plasmid DNA and PEG added. After this treatment the cells are washed and plated onto selective medium. This procedure is relatively simple to perform, but has the disadvantage that the efficiency of transformation is lower than spheroplast transformation and is strain dependent.

The third method of transformation is electroporation *(77)*. For transformation, plasmid DNA is added to a washed, osmotically-stabilized cell suspension that is exposed to an electric pulse. Afterwards the cells are plated on selective medium. This technique has the advantage of being simple, fast, and highly efficient.

Expression of Transformed Genes. Successful transformation of yeast does not necessarily guarantee that the target gene will be expressed. The level of gene expression is largely dependent upon the strength of the promoter being used. Promoters, which are non-coding DNA sequences that control gene expression, can be inserted before a target gene to enhance its expression *(78)*. Promoters of yeast genes encoding enzymes of the glycolytic pathway, such as phosphoglycerate kinase *(PGK1)* *(79)* and alcohol dehydrogenase *(ADH1)* *(80)*, have been used widely. These promoters are strong and are always active. Other promoters are tightly regulated and are controlled in response to external factors, such as particular sugars or ions. For example, the promoter for the gene encoding UDP-glucose-4-epimerase *(GAL10)*, an enzyme involved in galactose utilization, is activated when cells are grown in the presence of galactose *(81)*; and the gene encoding acid phosphatase *(PHO5)* is activated when cells are grown in medium containing low concentrations of inorganic phosphate *(82)*.

In addition to promoters, transcriptional terminators are required for high-level gene expression. Terminators are non-coding DNA sequences found at the end of genes that uniformly terminate gene transcription. Terminators from numerous genes including *ADH1* and *CYC1*, which encodes iso-1-cytochrome c, have been used to help express cloned genes *(78)*.

Various combinations of promoters and terminators have been inserted into plasmids, forming an expression cassette (Figure 8). Such plasmids are called expression vectors and have been engineered so they contain cloning sites located precisely between the promoter and terminator *(83)*. Thus, target genes can easily be inserted into an expression vector and placed under the control of a strong promoter for high-level gene expression.

In certain cases, it may be essential to have the product (a protein) secreted from the cell. This can be accomplished by inserting the target gene immediately downstream of a signal sequence *(78)*. Signal sequences are normally found at the beginning of genes whose products are secreted and encode a signal peptide, which directs the protein outside of the cell. The signal peptide protein is enzymatically removed to produce the mature protein during the process of secretion. Signal sequences from genes encoding α-factor *(MFα1)* and invertase *(SUC2)* are commonly used to direct the secretion of cloned gene products from yeast *(84)*.

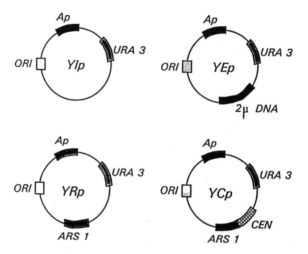

Figure 7.   Four basic types of plasmids for yeast transformation.

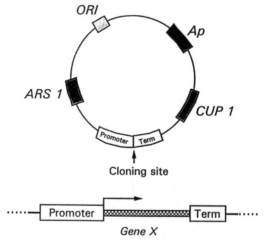

Figure 8.   Hypothetical plasmid containing an expression cassette.

Application Of Biotechnology To Brewers Yeast

Yeast is not only responsible for conversion of fermentable sugars to ethanol and carbon dioxide, but also for the production of unique flavor compounds, such as higher alcohols and esters.   The complexity of yeast function during fermentation, however, makes it difficult to change one attribute without affecting others and is a formidable challenge to the molecular biologist. Attempts to improve yeast performance fall into two broad categories:
1.   Reduced material costs
     - Ability to utilize normally non-fermentable carbohydrates
     - Ability to chillproof beer
     - Ability to degrade beta-glucans
2.   Increased production efficiency
     - Lower diacetyl production
     - Altered flocculation properties
     - Contamination resistance

Ability To Utilize Normally Non-Fermented Carbohydrates.   Brewing strains ferment a limited number of sugars, namely glucose, fructose, galactose, sucrose, maltose, and maltotriose.   Larger dextrins, such as maltotetraose, that are present in wort remain at the end of fermentation as unfermentable carbohydrates.

Over the past 20 years there has been a trend in the USA toward lighter, lower calorie beers, which makes it critical to have a high degree of conversion of dextrins to fermentable sugars during mashing.   Brewers have used a variety of techniques to produce highly fermentable worts, such as increasing mashing times or addition of fungal glucoamylase to either the mash or the fermentor.

An elegant solution for producing light beers that avoids the expense of buying fungal enzymes is to have yeast produce an amylase during fermentation, creating a specialty yeast for the production of light beers.   As early as 1971, a dextrin-utilizing yeast had been bred by crossing a lager yeast with *Saccharomyces diastaticus*, a wild yeast that produces a glucoamylase.   By repeated backcrossings with a lager yeast, a stable strain was produced that utilized dextrins and did not give the characteristic phenolic off-flavor associated with *S. diastaticus (85)*.   A similar hybrid yeast was produced by Russell *et al., (86)* and although they were able to construct a diploid strain that did not produce a phenolic-off flavor, it imparted undesirable winey and sulfury notes to the beer.

Because breeding of hybrid yeasts results in the incorporation of undesirable as well as desirable characteristics, geneticists have turned to genetic transformation, which allows the transfer and expression of single genes such as those that encode amylolytic enzymes.   A yeast episomal plasmid, containing the *S. diastaticus* glucoamylase gene, was used to transform a brewers yeast *(87)*.   In pilot plant trials this yeast utilized 30% of the wort dextrins; however, the fermentation rate was much slower than the parent strain *(88)*.   Furthermore, a normal fermentation rate could only be achieved by mixing the transformed strain with the parent strain, making it impractical for production brewing.

To address potentially unfavorable regulations for genetically-engineered yeasts, a lager yeast was transformed with a plasmid containing the glucoamylase gene of the yeast *S. diastaticus* in such a way that no bacterial DNA could be detected *(89)*. This dextrin-utilizing yeast fermented to a lower attenuation limit and gave a normal fermentation rate. However, after 60 generations only 12% of the cells produced glucoamylase, causing these workers to recommend only ten successive fermentations with this yeast.

The glucoamylase of the fungus *Aspergillus niger*, which is capable of superattenuating wort since it possesses debranching activity, has been integrated into the chromosome of a lager yeast *(90)*. This yeast was able to utilize approximately 50% of the wort dextrin with further utilization apparently being hampered by yeast flocculation. Although the beer produced had increased levels of esters and higher alcohols, it was slightly preferred to a control beer.

The glucoamylases from *A. niger* and *S. diastaticus* have the disadvantage of being thermostable. Both glucoamylases survive pasteurization and convert dextrins to glucose in the package, making the beer progressively sweeter. Once a strain producing a thermolabile glucoamylase is developed, however, much work must be done in brewery logistics to avoid mixing a dextrin-utilizing strain with a normally attenuating yeast.

Ability to Chill-Proof Beer. Before beer is packaged, it is chill-proofed to remove either proteins or polyphenols that combine and precipitate at cold temperatures to produce what is known as a chill-haze. Brewers use various methods to prevent chill-haze formation, including treatment of the matured beer with the proteolytic enzyme papain. Alternatively, a yeast could be engineered to produce a chill-proofing protease and so eliminate the cost and labor of this treatment. A secreted protease from a wild yeast has been cloned into a plasmid and used to transform brewer's yeast. Trial fermentations of a lager yeast transformed with this plasmid gave normal fermentations, but elevated diacetyl levels *(91)*. Although protease-producing yeast have the potential to eliminate chill-proofing, the type of protease must be carefully evaluated since prolonged contact times may have a detrimental effect on foam-active proteins.

Ability To Degrade Beta-Glucans. Beta-glucans are a family of polysaccharides composed of unbranched chains of ß-D-glucopyranose residues joined by (1-4) and (1-3) linkages. Generally, adjunct-containing worts are low in ß-glucans; however, use of poorly modified malt or barley in the mash will result in high levels of undegraded ß-glucans in the wort. As a consequence of the high ß-glucan content, these worts are viscous and difficult to lauter. In addition, beer made from high ß-glucan worts is difficult to filter, as well as being likely to produce insoluble gums and hazes.

Wort and beer ß-glucan levels can be reduced by adding a commercial ß-glucanase during mashing; however, to avoid this cost, brewer's yeasts have been engineered to secrete beta-glucanase into wort during fermentation. Early attempts to transform brewer's yeast with a ß-glucanase gene from the bacterium *Bacillus subtilis* were not successful, as only low amounts of ß-glucanase were produced *(92)*. Subsequently, expression of the ß-glucanase was improved by addition of a promoter sequence from the *MFα1* gene; however, very little was secreted into the beer *(93)*. Lancashire and Wilde *(94)* constructed a plasmid, which contained the ß-glucanase gene next to the signal sequence of the yeast α-factor gene to direct secretion of ß-glucanase into the wort. Laboratory fermentations with this yeast resulted in degradation of 33% of the wort ß-glucan.

In another approach, the genes encoding two ß-endoglucanases from the fungus *Trichoderma reesei* were inserted into an expression cassette using the *PGK1* promotor and *CUP1* as a selectable marker *(95)*. A yeast transformed by this plasmid degraded ß-glucans, although the transformant containing the genes integrated into the chromosome retained only 10% of the glucanolytic activity compared to a strain carrying the same genes on a multicopy plasmid *(96)*. The plasmid-containing strain was reported to be about 90% stable after seven fermentations and the beer was judged to be indistinguishable from beer produced by the untransformed strain.

Despite some successes in constructing a glucanolytic brewer's yeast, production of the ß-glucanase during the fermentation only has the potential to solve beer filterability problems and not wort filterability problems. Poor filterability in the brewhouse due to elevated levels of ß-glucan can still only be remedied by altering the grain bill, better malt modification, addition of glucanolytic enzymes to the mash, or by genetic engineering of the barley.

Lower Diacetyl Production. During isoleucine and valine synthesis, which occurs during alpha fermentation, excess acetohydroxyacids are excreted from the yeast into the beer and there are chemically converted to the vicinal diketones, diacetyl, and pentanedione (Figure 9). Diacetyl imparts a buttery or butterscotch flavor to the beer that is undesirable. Therefore, beers are lagered or matured for days or weeks primarily to enable yeast to reduce diacetyl first to acetoin and then to the tasteless 2,3-butanediol. Here we will discuss primarily methods to reduce diacetyl, although pentanedione should also be similarly affected.

Reduction of beer lager time will improve brewing efficiencies, since it will result in faster throughput and less capital investment in tankage. These advantages have been recognized by brewers, and a number of different strategies have been used reduce the amount of diacetyl produced by brewer's yeast and so shorten the maturation time. These strategies can be categorized as follows:

1. Reducing levels of acetohydroxyacid synthetic enzymes.
2. Increasing pathway enzyme levels after acetohydroxyacid synthesis.
3. Engineering a yeast to produce α-acetolactate decarboxylase.
4. Selecting a yeast with altered acetohydroxyacid synthase activity.

The first approach is to decrease acetohydroxyacid production, which theoretically can be achieved by mutating the *ILV2* gene, which produces acetohydroxyacid synthase (AHAS). Although this has been accomplished in laboratory yeast strains, it was accompanied by a loss in fermentation vigor *(97)*,

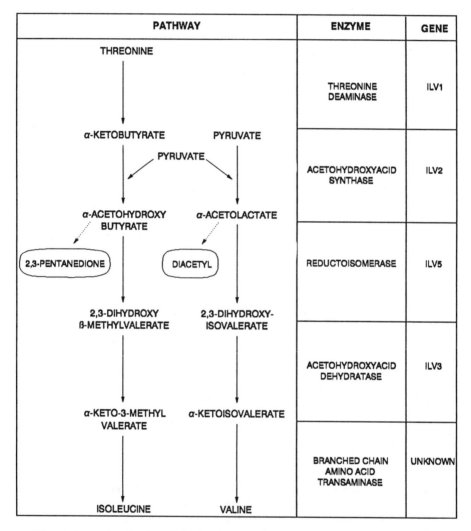

Figure 9.   Synthesis of isoleucine, valine, and vicinal diketones by brewer's yeast.

which is not surprising as reduced synthesis of the valine and isoleucine would be expected to affect normal yeast metabolism.

The next approach has been to increase the flux down the isoleucine/valine pathways. Acetohydroxyacid reductoisomerase (AHAR) levels have been shown to be a limiting step in the synthesis of isoleucine and valine, causing an accumulation of α-acetolactate. The limiting effect of AHAR was confirmed when the *ILV5* gene was introduced into brewer's yeast on a multicopy plasmid *(98)*. These strains showed a 70% reduction in vicinal diketone production. Since it is difficult to stabilize plasmids, particularly in industrial yeasts, a plasmid containing *ILV5* has been successfully integrated into chromosome VIII *(99)*, although data on the stability and performance of these transformants has not been published.

In the 1980's, researchers found that addition of the bacterial enzyme α-acetolactate decarboxylase (ALDC) to beer during primary fermentation reduced diacetyl below the flavor threshold by the end of primary fermentation, eliminating the need for lagering. ALDC converts α-acetolactate directly to acetoin, eliminating the need for chemical conversion of acetolactate to diacetyl. An elegant application of this approach has been to integrate the ALDC of the entire bacteria, *Enterobacter aerogenes* or *Klebsiella terrigena* into brewer's yeast using *PGK1* or *ADH1* promotors and *CUP1* as a selectable marker *(100-102)*. In 50-liter fermentation trials of these transformed strains, no detectable diacetyl was present at the end of the primary fermentation, and total lager beer production time was reduced from three to one week *(102)*. Although the fusel alcohol levels were slightly elevated in beer from the transformed strain, it was judged to taste similar to beer produced with the parent yeast.

A drawback to transformed yeast strains is that they require regulatory approval and can be construed as non-traditional by the consumer. A significant reduction in the diacetyl production has been achieved, however, by selection of spontaneous mutants from brewers yeast cultures using resistance to the herbicide, sulfometuron methyl *(103)*. Sulfometuron methyl resistant strains have a partial inactivation of acetohydroxyacid synthetase so they produce 50% less diacetyl than the parent strain *(67)*. Recently Sigsgaard and Hjortshoj *(104)* reported that they had constructed a yeast using traditional genetic techniques, such as sporulation, mutation, mating, and selection that produced a beer with 10% of the normal diacetyl at the end of primary fermentation. This yeast apparently has performed well in brewery trials.

Altered Flocculation Characteristics.   A key property of yeasts is that they aggregate and then sediment toward the end of fermentation. This property, known as flocculation, is of great benefit to the brewer as it allows a lager yeast to be harvested from the bottom of the fermentor and means that only small quantities of yeast must be filtered from the matured beer. While the advent of the centrifuge has made flocculation less significant to some brewers, it is an important and much researched characteristic *(105)*. Although the exact mechanism and components responsible for flocculation are still unknown, eleven genes have identified *(105)*. Most of the research efforts have been devoted to the dominant genes *FLO1*, *2*, *4*, *5*, and *8*, which promote flocculation. The *FLO1* gene has been cloned and introduced on a plasmid into non-flocculent brewer's yeasts *(106)*. All the non-flocculent yeasts that were transformed with *FLO1* became flocculent. Nevertheless, these transformants were quite unstable, with only 37% of the cells remaining flocculent after two generations. Recently Vezinhet *et al.* *(57)*, introduced chromosome I (which contains *FLO5*) from a flocculent strain to a wine yeast by cytoduction. Flocculent clones were isolated although many were unstable. Furthermore, the clones that were stable gave slower fermentations.

Contamination Resistance. In every brewery operation there is always the risk of contamination from wild (non-brewing) yeasts. Many wild yeasts contain a killer-factor that encodes the toxin zymocin and imparts immunity to the killer toxin. Brewer's yeasts do not carry the killer factor and so will be killed if a killer yeast contaminates the culture. Brewers yeasts, however, can be transformed into killer strains by cytoduction *(107, 108)*. Unfortunately, many of the killer brewers yeasts that have been isolated are poor fermenters, which has been attributed to the introduction of mitochondria from the killer wild yeast strain.

While a yeast that can kill contaminating yeast is an attractive proposition, it would only kill non-toxin carrying yeast; wild yeast carrying the killer factor would be unaffected. Furthermore, the yeast would still be susceptible from bacterial contamination, which has prompted other researchers to suggest that strains should produce bacteriocins as well as zymocins. Creation of an antimicrobial yeast has been accomplished by Sasaki *et al.* *(109)*, who mated a respiratory-deficient strain with antibacterial properties with a killer yeast, and then fused this hybrid strain with a brewers yeast (Figure 10). The resulting strain possessed antibacterial properties as well as the yeast killer factor. In fermentation trials, however, the brewers yeast killer/antibacterial hybrid fermented slower than the brewing strain. Once the technical problems have been solved, the introduction of a killer strain into a brewery presents a number of logistical problems, since accidental mixing would prove fatal for a normal brewers yeast. Therefore, a successful introduction of a killer brewers yeast into a brewery may require that all the brewery yeasts are killer strains.

Future Development in Brewing

Biotechnology clearly offers an exciting array of tools that will improve brewing ingredients. Nevertheless, a number of technical challenges lie ahead. While stable transformants have been produced in yeast, they have yet to become a reality in barley and hop. Still more work needs to be done on the basic biochemistry of agricultural materials and yeast to guide the geneticist in selecting appropriate genes. Careful evaluation of improvements to the brewing ingredients must be done to ensure that the modification has a practical value in simplifying production. A modified ingredient which solves one problem, but creates another is unlikely to be embraced by the brewer.

Other areas that need addressing fall into the categories of regulatory, legal, and consumer affairs. Government regulation concerning the approval of

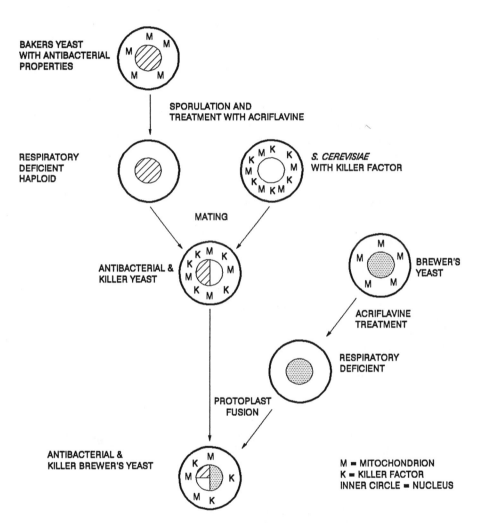

Figure 10. Construction of antibacterial and killer brewer's yeast (adapted from ref. 109).

bioengineered foods has been unclear at best. Recently, however, the Ministry of Agriculture Food and Fisheries in the UK approved the use of a baker's yeast that had been genetically engineered to improve maltose fermentation. As the genetic transfer was within a single yeast species, it was considered safe. In the USA the Food and Drug Administration recently announced guidelines for bioengineered foods that eliminate formal pre-market reviews of most genetically engineered foods (110). Environmental groups have voiced opposition to this plan, however, and it remains unclear how and when new products such as Calgene's slow-softening Flavr Savr tomato, will be introduced into the marketplace (111).

The intellectual property rights for many of the basic techniques in bioengineering have been patented, so products developed using these techniques are subject to royalty payments. Therefore, brewers will have to make a cost benefit analysis that includes royalty payments on any modified products.

A final, but by no means less important issue, is that of consumer acceptance. While consumers expect therapeutic drugs to be made by the latest technologies, beer is consumed with the expectation that natural ingredients and traditional methods are used. Because of these constraints it is unlikely that breweries will ever be in the vanguard of biotechnological developments.

Acknowledgments

The authors thank Anheuser-Busch Companies, Inc., and Mr. Raymond E. Goff for permission to publish this review. We also thank Ms. Patti Abbott for her patience in formatting and editing this document and her excellent assistance, along with Ms. Pat Erdmann, in developing the figures.

References

1. Jensen, N. F. *Plant Breeding Methodology*: John Wiley & Sons, Inc.: New York, NY, 1988.
2. Wych, R. D.; Rasmusson, D. C. *Crop Science* **1983**, *23*, 1037-1040.
3. Heale, J. B.; Legg, T.; Connell, S. In *Biotechnology In Agriculture And Forestry*; Bajaj, Y. P. S., Ed.; Medicinal and Aromatic Plants II; Springer-Verlag: Berlin, 1989, Vol. 7; pp 264-285.
4. Sasaki, M.; Shikata, E. *Rev. Plant Prot. Res.* **1980**, *13*, 97-113.
5. Puchta, H.; Ramm, K.; Sanger, H. L. *Nucleic Acids Res.* **1988**, *16*, 4197-4216.
6. Gunn, R.; Desmond, B.; Adams, T. *Brewers Digest* **1991**, *66*, 18-20.
7. Schlosser, S. *Seed World* **1989**, *127*, 26-32.
8. Dallas, J. F. *Proc. Natl. Acad. Sci. USA* **1988**, *85*, 6831-6835.
9. Welsh, J.; *et al. Theor. Appl. Genet.* **1991**, *82*, 473-476.
10. Hagedorn, A. *Wall Street Journal* **1990** (March 5, 1990), B1.
11. Siebel, J.; Pauls, K. P. *Theor. Appl. Genet.* **1989**, *78*, 473-479.
12. Picard, E. *Genome* **1989**, *31*, 1005-1013.
13. Tanksley, S. D.; *et al. Biotechnology*, 7, 257-264.
14. Paterson, A. H.; *et al. Nature* **1988**, *335*, 721-726.

15. Lee, M.; et al. *Crop Science* **1989**, *29*, 1067-1071.
16. Williams, J. G. K.; et al. *Nucleic Acids Res.* **1990**, *18*, 6531-6535.
17. Martin, G. B.; Williams, J. G. K.; Tanksley, S. D. *Proc. Natl. Acad. Sci. USA* **1991**, *88*, 2336-2340.
18. Kleinhofs, A.; et al. *Theor. Appl. Genet.* in press.
19. Hockett, E. A.; Nilan, R. A. In *Barley*; Rasmusson, D. C., Ed.; Agronomy; American Society of Agronomy: Madison, WI, 1985, Vol. 26; pp 187-230.
20. Von Wettstein, D.; et al. *MBAA Tech. Quart.* **1980**, *17*, 16-23.
21. Scowcroft, W. R.; et al. In *Plant Tissue and Cell Culture*; Green, C. E.; et al., Eds.; Plant Biology; Alan R. Liss, Inc.: New York, NY, 1987, Vol. 3; pp 275-286.
22. Xie, Q. J.; et al. *Crop Science* **1992**, *32*, 507.
23. Duncan, R. R.; et al. *Crop Science* **1991**, *31*, 242-244.
24. Miller, D. R.; et al. *Crop Science* **1992**, *32*, 324-327.
25. Gengenback, B. G.; Green, G. E.; Donovan, C. M. *Proc. Natl. Acad. Science USA* **1977**, *74*, 5113-5117.
26. Jahne, A.; et al. *Theor. Appl. Genet.* **1991**, *82*, 74-80.
27. Datta, S. D.; et al. *Biotechnology* **1990**, *8*, 736-740.
28. Powell, A.; Gordon, M. P. In *Molecular Biology*; Marcus, A., Ed.; The Biochemistry of Plants; Academic Press, Inc.: San Diego, CA, 1989, Vol. 15; pp 617-651.
29. Fraley, R. T.; Rogers, S. G.; Horsch, R. B. *CRC Crit. Rev. Plant Science* **1986**, *4*, 1-46.
30. Shimamoto, K.; et al. *Nature* **1989**, *338*, 274-276.
31. Fromm, M. E.; et al. *Biotechnology* **1990**, *8*, 833-839.
32. Vasil, V.; et al. *Biotechnology* **1992**, *10*, 667-674.
33. Christou, P.; Ford, T. L.; Kofron, M. *Trends in Biotechnology* **1992**, *10*, 239-246.
34. D'Halluin, K.; et al. *Plant Cell* **1992**, *4*, 1495-1505.
35. Okamuro, J. K.; Goldberg, R. B. In *Molecular Biology*; Marcus, A., Ed.; The Biochemistry of Plants; Academic Press, Inc.: San Diego, CA, 1989, Vol. 15; pp 2-82.
36. Rodermel, S. R.; Abbott, M. S.; Bogorad, L. *Cell* **1988**, *55*, 673-681.
37. Spencer, J. F. T.; Spencer, D. M. *Ann. Rev. Microbiol.* **1983**, *37*, 121-142.
38. Johnston, J. R. In *Yeast Technology*; Spencer, J. F. T.; Spencer, D. M., Eds.; Springer Verlag: Berlin, Heidelberg, 1990; pp 55-104.
39. Casey, G. P. In *Yeast Strain Selection*; Panchal, C. J., Eds.; Bioprocess Technology; Marcel Dekker, Inc.: New York and Basel, 1988, Vol. 8; pp 65-111.
40. Olson, M. V. In *Genome Dynamics, Protein Synthesis and Energetics*; Broach, J. R.; Pringle, J. R.; Jones E. W., Eds.; The Molecular and Cellular Biology of the Yeast Saccharomyces; Cold Spring Harbor Laboratory Press: Cold Spring Harbor, NY, 1991, Vol. 1; pp 1-39.
41. Newlon, C. S. In *The Yeasts*; Rose, A. H.; Harrison, J. S., Eds.; Academic Press, Inc.: San Diego, CA, 1989, Vol. 3; pp 57-115.
42. Johnston, J. R.; Oberman, H. In *Progress in Industrial Microbiology*; Bull, M. J., Eds.: Elsevier: Amsterdam, 1979, pp 151-205.

43. Broach, J. R.; Volkert, F. C. In *Genome Dynamics, Protein Synthesis, and Energetics*; Broach, J. R.; Pringle, J. R.; Jones, E. W., Eds.; The Molecular and Cellular Biology of the Yeast Saccharomyces; Cold Spring Harbor Laboratory Press: Cold Spring Harbor, NY, 1991, Vol. 1; pp 297-331.

44. Pon, L.; Schatz, G. In *Genome Dynamics, Protein Synthesis, and Energetics*; Broach, J. R.; Pringle, J. R.; Jones, E. W., Eds.; The Molecular and Cellular Biology of the Yeast Saccharomyces; Cold Spring Harbor Laboratory Press: Cold Spring Harbor, NY, 1991, Vol. 1; pp 333-406.

45. Wickner, R. B. In *Genome Dynamics, Protein Synthesis, and Energetics*; Broach, J. R.; Pringle, J. R.; Jones, E. W., Eds.; The Molecular and Cellular Biology of the Yeast Saccharomyces; Cold Spring Harbor Laboratory Press: Cold Spring Harbor, NY, 1991, Vol. 1; pp 263-296.

46. Pringle, J. R.; Hartwell, L. H. In *Life Cycle and Inheritance*; Strathern, J. N.; Jones, E. W.; Broach, J. R., Eds.; The Molecular Biology of the Yeast Saccharomyces; Cold Spring Harbor Laboratory Press: Cold Spring Harbor, NY, 1981; pp 97-142.

47. Thorner, J. In *Life Cycle and Inheritance*; Strathern, J. N.; Jones, E. W.; Broach, J. R., Eds.; The Molecular Biology of the Yeast Saccharomyces; Cold Spring Harbor Laboratory Press: Cold Spring Harbor, NY, 1981; pp 143-180.

48. Esposito, R. E.; Klapholz, S. In *Life Cycle and Inheritance*; Strathern, J. N.; Jones, E. W.; Broach, J. R., Eds.; The Molecular Biology of the Yeast Saccharomyces; Cold Spring Harbor Laboratory Press: Cold Spring Harbor, NY, 1981; pp 211-287.

49. Gjermansen, C.; Sigsgaard, P. *Carlsberg Res. Commun.* **1981**, *46*, 1-11.

50. Bilinski, C. A.; Russell, I.; Stewart, G. G. *Proc. 21st Cong. Eur. Brew. Conv., Madrid* **1987**, 497-504.

51. Gunge, N.; Nakatomi, Y. *Genetics* **1972**, *70*, 41-58.

52. Spencer, J. F. T.; Spencer, D. M. *J. Inst. Brew.* **1977**, *83*, 287-289.

53. Tubb, R. S.; *et al. Proc. 18th Cong. Eur. Brew. Conv., Copenhagen* **1981**, 487-491.

54. Conde, J.; Fink, G. R. *Proc. Natl. Acad. Sci.* **1976**, *73*, 3651-3655.

55. Dutcher, S. K. *Mol. Cell Biol.* **1981**, *1*, 246-253.

56. Nilsson-Tillgren, T.; *et al. Carlsberg Res. Commun.* **1981**, *46*, 65-76.

57. Vezhinet, F.; *et al. J. Inst. Brew.* **1992**, *98*, 315-320.

58. van Solingen, P.; van der Plaat, J. B. *J. Bacteriol.* **1977**, *130*, 946-947.

59. Spencer, D. M.; Reynolds, N.; Spencer, J. F. T. In *Yeast Technology*; Spencer, D. M., Eds.; Springer Verlag: Berlin, Heidelberg, 1990; pp 348-354.

60. Christensen, B. E. *Carlsberg Res. Commun.* **1979**, *44*, 225-233.

61. Spencer, J. F. T.; *et al. Curr. Genet.* **1983**, 7, 159-164.

62. Molzahn, S. W. *J. Am. Soc. Brew. Chem.* **1977**, *35*, 54-59.

63. Oakley-Gutowski, K. M.; Hawthorne, D. B.; Kavanagh, T. E. *J. Am. Soc. Brew. Chem.* **1992**, *50*, 48-52.

64. Hall, J. F. *J. Inst. Brew.* **1971**, 77, 513-517.

65. Sigsgaard, P.; Pedersen, M. B. *Eur. Brew. Conv. Micro Group Bulletin, Copenhagen* **1990**, 43-59.

66. Kielland-Brandt, M. C.; *et al. Yeast* **1988**, *4 (special issue)*, S470.
67. Falco, S. C.; Dumas, K. S. *Genetics* **1985**, *109*, 21-35.
68. Boguslawski, G. In *Gene Manipulations in Fungi*; Bennett, J. W.; Lasure, L. L., Eds.; Academic Press, Inc.: Orlando, 1985; pp 161-195.
69. Kessler, D. A.; *et al. Science* **1992**, *256*, 1747-1749.
70. Henderson, R. C. A.; Cox, B. S.; Tubb, R. *Curr. Genet.* **1985**, *9*, 133-138.
71. Casey, G. P.; Xiao, W.; Rank, G. H. *J. Inst. Brew.* **1988**, *94*, 93-97.
72. Webster, T. D.; Dickson, R. C. *Gene* **1983**, *26*, 243-252.
73. Gritz, L.; Davies, J. *Gene* **1983**, *25*, 179-188.
74. Hinnen, A.; Hicks, J. B.; Fink, G. R. *Proc. Natl. Acad. Sci.* **1978**, *75*, 1929-1933.
75. Beggs, J. D. *Nature* **1978**, *275*, 104-109.
76. Ito, H.; *et al. J. Bacteriol.* **1983**, *153*, 163-168.
77. Becker, D. M.; Guarente, L. *Meth. Enzymol.* **1991**, *194*, 182-187.
78. Romanos, M. A.; Scorer, C. A.; Clare, J. J. *Yeast* **1992**, *8*, 423-488.
79. Hitzeman, R. A.; *et al. Science* **1983**, *219*, 620-625.
80. Hitzeman, R. A.; *et al. Nature* **1981**, *293*, 717-723.
81. Johnston, M. *Microbiol. Rev.* **1987**, *51*, 458-476.
82. Kramer, R. A.; *et al. Proc. Natl. Acad. Sci.* **1984**, *81*, 367-370.
83. Schneider, J. C.; Guarente, L. *Meth. Enzymol.* **1991**, *194*, 373-388.
84. Moir, D. T.; Davidow, L. S. *Meth. Enzymol.* **1991**, *194*, 491-507.
85. Emeis, C. C. *Am. Soc. Brew. Chem. Proc.* **1971**, *29*, 58-62.
86. Russel, I.; Hancock, I. F.; Stewart, G. G. *J. Am. Soc. Brew. Chem.* **1983**, *41*, 45-51.
87. Meaden, P. G.; Tubb, R. S. *Proc. 20th Cong. Eur. Brew. Conv. Helsinki* **1985**, 219-226.
88. Perry, C.; Meaden, P. *J. Inst. Brew.* **1988**, *94*, 64-67.
89. Vakeria, D.; Hinchliffe, E. *Proc. 22nd Cong. Eur. Brew. Conv. Zurich* **1989**, 475-482.
90. Gopal, C. V.; Hammond, J. R. M. *Proc. 5th Intl. Brew. Tech. Conf. Harrogate* **1992**, 297-306.
91. Young, T. W.; Hosford, E. A. *Proc. 21st Cong. Eur. Brew. Conv. Madrid* **1987**, 521-528.
92. Cantwell, B. A.; McConnell, D. J. *Gene* **1983**, *23*, 211-219.
93. Lancashire, W. E.; Wilde, R. J. *Proc. 21st Cong. Eur. Brew. Conv. Madrid* **1987**, 513-520.
94. Cantwell, B. A.; *et al. Curr. Genet.* **1986**, *11*, 65-70.
95. Penttila, M. E.; *et al. Yeast* **1987**, *3*, 175-185.
96. Ramos-Jeunehomme, C.; *et al. Proc. 20th Cong. Eur. Brew. Conv. Helsinki* **1985**, 283-290.
97. Enari, T. M.; *et al. Proc. 21st Cong. Eur. Brew. Conv. Madrid* **1987**, 529-536.
98. Villanueba, K. D.; Goosens, E.; Masschelein, C. A. *J. Am. Soc. Brew. Chem.* **1990**, *48*, 111-114.
99. Goosens, E.; *et al. Proc. 23rd Cong. Eur. Brew. Conv. Lisbon* **1991**, 239-296.
100. Penttila, M. E.; *et al. Yeast* **1988**, *4*, S473.

101. Suihko, M. L.; *et al. Proc. 22nd Cong. Eur. Brew. Conv. Zurich* **1989**, 483-490.
102. Blomqvist, K.; *et al. App. Environ. Micro.* **1991**, *S7*, 2796-2803.
103. Kielland-Brandt, M. C.; *et al. Yeast* **1988**, *4*, S470.
104. Sigsgaard, P.; Hjortshoj, B. *Proc. 5th Intl. Brew. Tech. Conv. Harrogate* **1992**, 119-127.
105. Speers, R. A.; *et al. J. Inst. Brew.* **1992**, *98*, 293-300.
106. Watari, J.; *et al. Proc. 23rd Cong. Eur. Brew. Conv. Lisbon* **1991**, 297-304.
107. Young, T. W. *J. Am. Soc. Brew. Chem.* **1983**, *41*, 1-4.
108. Hammond, J. R. M.; Eckersley, K. W. *J. Inst. Brew.* **1984**, *90*, 167-177.
109. Sasaki, T.; *et al. J. Am. Soc. Brew. Chem.* **1984**, 164-166.
110. Kessler, D. A.; *et al. Science* **1992**, *256*, 1747-1832.
111. Hamilton, J. O. C., Ellis, J. E.; *Business Week* **1992**, *12/14/92*, 98-101.

RECEIVED May 3, 1993

Chapter 10

Use of Enzymes in Wine Making and Grape Processing

Technological Advances

Peter F. H. Plank and James B. Zent

Genencor International, Inc., 180 Kimball Way, South San Francisco, CA 94080

The use of biotechnology in winemaking and grape processing has extended beyond the concept of yeast and malolactic fermentation technology. Food grade industrial enzymes offer significant quantitative and qualitative processing improvements to both the winemaker and the grape processor. These combined factors result in overall economic benefits to the processor. Food grade enzymes offer quantitative benefits in the form of increased free run and press juice yields. The qualitative benefits are derived from improved color extraction in red grape varieties, increased fermentable sugar recovery, and improvements in the aging process of wines, *i.e.* flavor enhancement or modification. Processing benefits achieved using enzymes allow for improved handling of difficult-to-press grape varieties and faster lees settling rates, as well as reduced must viscosity for improved downstream processing, *e.g.* filtration. Future biotechnological advances will facilitate increased use of enzymes in these and other potential applications.

Wine is simple proof that God loves us, and a constant reminder that he intends for us to be happy.

— Benjamin Franklin

People have always been fascinated with fermentation and its end products. Fermented products such as beer and wine were considered more than just nutritional beverages; they were touted as having medicinal properties as well.

Throughout history, many of the basic principles of fermentation were "worked out" serendipitously. It was not until 200 years ago that the revolution in fermentation began. What initially was considered an art had evolved into a science.

The early fermentation work of Lavoisier, Pasteur, Traube, Kuhne and others in the 19th century was further developed by the likes of Neuberg, Harden, Warburg and others in the years that followed (*1-3*). All of these pioneers helped develop fermentation technology, and thus winemaking, leading both into the realm of the sciences. Even today, however, we cannot explain the chemistry of winemaking in a completely scientific manner. While we continue to make advances in these areas, much work remains to be carried out in order to determine the composition of grapes and wine, the factors influencing fermentation, aging, quality, and the like.

0097–6156/93/0536–0181$06.00/0

The History of Enzyme Usage in Winemaking

The first endogenous enzyme used in winemaking was a pectinase. During the 1940's, the use of pectinase for the clarification of apple juice was fairly common. Early attempts to clarify grape juice, or *must*, with these commercial pectinases resulted in insufficient clarification and/or high methanol production.

In the early 1950's, Rohm & Haas commercialized PECTINOL 59L, the first pectinase developed specifically for winemaking. As analytical assay methods improved, the pectolytic activities in this new pectinase were identified as being predominantly pectin lyase (PL). Previous pectinases were composed primarily of polygalacturonase (PG) and pectin methyl esterase (PME) activity. It was subsequently determined that pectin lyase and polygalacturonase have different reaction mechanisms for pectin degradation. Polygalacturonase requires pectin methyl esterase in order to hydrolyze pectin, since polygalacturonase only recognizes demethoxylated pectin as a substrate. In contrast, pectin lyase only recognizes methoxylated pectin as a substrate. PECTINOL 59L, being predominantly pectin lyase, naturally produced very little methanol. PECTINOL 59L also contained additional carbohydrase side activities, such as arabinogalactanase, which also contributed to its efficacy in grape must clarification.

In the mid-1950's, Rohm & Haas continued development of pectinases for winemaking with emphasis being placed on: *1)* improving processing and/or quality, and *2)* anthocyanin hydrolysis. This work resulted in the commercialization of additional pectinases in the 1960's for increasing free run and total juice yield, pressing efficiency, color extraction, filtration, etc., as well as for color removal or anthocyanin hydrolysis.

Obviously these pectinases contained significant levels of side activities, which contributed to their improved performance. In pectinases for processing and/or quality improvements, the side activities were identified as being carbohydrases; in particular, arabinogalactanases (4), beta-glucanases, dextranases and cellulases. In pectinases for anthocyanin hydrolysis, the important side activities are the glycosidases, for example, glucosidase, galactosidase, arabinosidase and rhamnosidase.

One of the interesting influences associated with this new breed of pectinases regarding wine quality was the effect on flavor development and aroma modification. These effects were far greater with the anthocyanase-containing pectinases than with the pectinases developed specifically for processing improvements. In the mid-1980's, research on wine flavor development elucidated the reaction mechanisms involved in this phenomenon. The chemical structures of these flavor precursors were also characterized at this time. These new pectinases contained sufficient amounts of glycosidase side activities which were specific for these flavor and aroma precursors. These glycosidases hydrolyze the glycosidic portion of the flavor and aroma precursors, similar to the hydrolysis mechanism associated with anthocyanins. [Both anthocyanin hydrolysis and flavor and aroma development will be covered more thoroughly later.]

Pectinase usage in the wine industry became an accepted practice in the 1970's. During this period, further evaluation of additional enzyme activities, such as cellulase, protease, and glucose oxidase, were conducted. These activities were attempts to improve processing and to reduce hazing and browning. The results generally demonstrated little benefit, and often showed adverse side effects.

In the mid-1980's, the development of macerating enzymes for winemaking began, again due to the earlier success with these enzymes in the apple juice industry. Today, these macerating enzymes are rapidly replacing pectinases in winemaking around the world, due to their additional benefits. These benefits will be covered in more detail later.

Composition of Wine Grapes

The chemical composition of wine grapes *(Vitus vinifera)* and of wine has been the subject of considerable research. The variability associated with the composition of the different varieties is well documented. Only those chemical and structural components of specific interest to this article will be reviewed here. These include endogenous enzymes, pectin, cellulose, glucan, hemicellulose, protein, lignin, polyphenolics and related compounds, and flavor precursors.

Enzymes endogenously present in grapes (*1*) include pectinase, polyphenol oxidase (PPO), peroxidase, catalase, tannase, invertase, ascorbase, dehydratrase, esterase, protease, and the glycosidases (*5-8*). PPO has received considerable attention due to its effect on wine quality and the reduction of browning.

Recently, glycosidases in grapes have been receiving considerable attention due to their role in flavor development (*5-8*). The other endogenous enzymes have received lesser attention due to the minor role they play in wine quality.

The levels of all of these enzymes vary greatly among wine grapes, from being nearly nonexistent in some varieties to being present in large amounts in others (*8*). The presence and role of these endogenous enzymes in grapes has been reviewed by Amerine and Joslyn (*1*).

The cell walls of grapes are composed primarily of a mixture of complex carbohydrates such as pectin, cellulose, and hemicellulose. Non-carbohydrate based compounds such as lignin and protein are also present. The levels of these constituents vary greatly among the many different wine grape varieties. These compounds are of particular interest for the application of hydrolytic enzymes for processing, yield, and quality improvements in winemaking.

The proteins in grapes are quite complex, and their exact structure are still unknown. The level of protein present, as well as its structure, appears to vary significantly among varieties. Yeast can also contribute to the total protein content of the wine during fermentation (*3*). This area has received considerable attention due to the problems associated with protein-induced hazing of wine.

Research conducted on the role of lignin in winemaking shows that lignin is present in trace amounts in grapes. The major source of lignin found in wine is the oak during barrel aging (*1, 9*). Lignin undergoes hydrolysis during the aging process of wine and brandy. The hydrolysis products of this reaction can undergo further oxidation, generating compounds which are important to the aging characteristics of wine.

Constituents of grapes, other than structural compounds, that are of particular importance include anthocyanins, tannins, flavanoids, proanthocyanidins, and other polyphenolics, as well as glycosides of terpenols, norisoprenoids, and shikimic acid derivatives (*10, 11*).

Anthocyanins, which are very reactive compounds, are of major importance in the color stability of red wines. The term *tannin* is used to describe two distinctly different classes of compounds. As defined originally, "tannin" referred to the "hydrolyzable tannins" which are composed of one molecule of glucose and many molecules of gallic acid. There is some evidence that these compounds may be present in some grape varieties, as well as in grape seeds. It is generally believed that the occurrence of hydrolyzable tannins in wine grapes is low. Tannins are present at significant levels in oak, and are considered quite important to the development of wine body, flavor and mouthfeel. Commercial tannins (tannic acid) which can be added to wine are of this class.

A second class of compounds often referred to as "tannins" are proanthocyanidins. This class consists of condensed polymers, derived mostly from flavan-3,4-diol (procyanidins) and 3-flavanol (catechins). Proanthocyanidins are inappropriately referred to as "condensed tannins". These are the main "tannins" in grapes, and they are very important to wine flavor and structure, *i.e.*, balance, body and mouthfeel. They can also affect wine quality due to their involvement in browning reactions (*12*) and interactions with anthocyanins. Both classes of so-called tannins are important due to their role in wine color, flavor, body and mouthfeel (Rohm & Haas, Genencor International Inc., unpublished data).

Commercially Available Pectinases

Traditionally, pectinases have been used to treat grapes in order to improve processing, either through increases in free run and total juice yield, improved pressability, or reduction of must viscosity by degrading pectin. Pectin (Figure 1) is a complex polymer composed of a galacturonic acid backbone with varying degrees of methylation. There are three distinct classes of pectic substances: *protopectin*, a water insoluble component; *pectinic acid* or *high methoxyl*, containing pectins and pectates; and *pectic acids* or *low methoxyl*, containing pectins and pectates. Commercially available pectinases are typically mixtures of three principal pectolytic activities: pectin methyl esterase (PME), pectin lyase or transeliminase (PL), and polygalacturonase (PG). Each of these pectolytic activities recognizes and acts upon a different portion of the pectin molecule.

Pectin methyl esterase recognizes and degrades the carboxymethyl groups of polygalacturonic acid (PGA), releasing methanol into the must (Figure 2). As a result, pectin methyl esterase has a deleterious effect on the quality of both red and white wines which are fermented on the skins for extended periods of time. Pectin lyase also recognizes the carboxymethyl group of polygalacturonic acid; however, its activity will result in the elimination of water and formation of a conjugated enone system (Figure 3). On the other hand, polygalacturonase recognizes the free carboxylic acid residues of polygalacturonic acid and results in the hydrolysis of the alpha-1,4 bonds, releasing galacturonic acid into the must (Figure 2). Both pectin lyase and polygalacturonase are very effective at reducing must viscosity by degrading the polygalacturonic acid in an endo manner. The ideal commercial pectolytic preparation would be a product which contains mostly pectin lyase and polygalacturonase activities. Pectin methyl esterase content should be minimal, preferably nonexistent, in the final formulation.

Technological Developments in Macerating Enzymes

As technology progressed, it was determined that the structure of the grape was not only composed of pectin (*13-14*). The typical amount of pectin present in the grape berry is dependent on a number of factors including the grape variety, its degree of maturity, soil, crop yield, and post-harvest handling. The pectin content has been reported to be in the range of 0.06% to 0.2% (*15-16*). In addition to pectin, there are a number of other structural components present which are not degraded by typical pectolytic preparations. These other components are colloidal in nature and are composed of polysaccharides or hemicelluloses, proteins and polyphenolics (*17-18*). When the grapes are infected with *Botrytis cinerea* the presence of beta-glucan has been observed (*19*). Colloids have been reported to be present at the level of 167 to 324 mg/L in white wines and 960 mg/L in red wine.

In order to improve processing further, many enzyme manufacturers began the development of macerating enzymes to address these structural components. The fermentation technology used to develop macerating enzymes is based on the early work in pectolytic fungal technology. These macerating enzyme fermentations take into account the structural components of the berry being processed. The resultant macerating enzyme fermentation product contains the relevant pectolytic, cellulolytic, and hemicellulolytic activities required to process the berry. The end results achievable using macerating enzyme formulations are readily seen in three main areas: *1*) improvements in processing; *2*) improvements in yield (both free run and pressed juice fractions); and *3*) improvements in the juice and wine quality. We shall address each of these topics.

Yield Improvements. There are a number of yield benefits associated with the use of macerating enzymes (ME). Increases in the level of free run juice yields of 5% to 15% have been observed, depending on the type of process. Both press juice and total juice yield recoveries improved through the use of macerating enzymes. Improvements of 5% to 15% in these parameters have been observed on a regular basis. The benefits associated with the use of macerating enzymes can be readily observed when one considers the high cost of some grape varieties being processed and the end use of the must. Macerating enzymes also improved lees compaction during clarification, resulting in a 5% to 15% increase in the clears volume. A direct benefit associated with this is a corresponding 10% to 30% decrease in filtration losses (Genencor International, Inc., unpublished data).

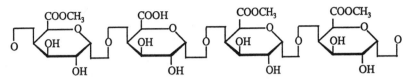

Figure 1: Structure of Pectin (Polygalacturonic Acid)

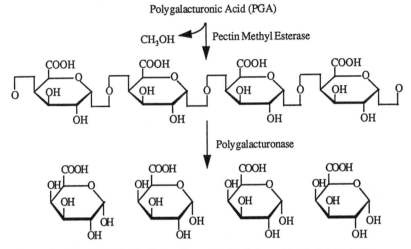

Figure 2: Pectin Hydrolysis Products Derived from Pectin Methyl Esterase and Polygalacturonase Activity

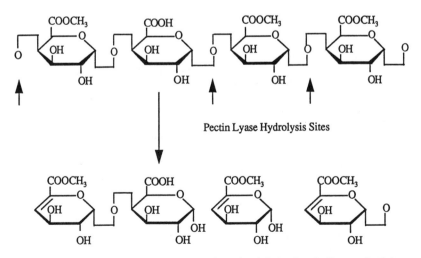

Figure 3: Pectin Hydrolysis Products Derived from Pectin Lyase Activity

Processing Improvements. The major processing benefits associated with the use of macerating enzymes are *1*) improved handling of "difficul to press' grape varieties, such as slip-skin varieties (for example, Muscat); *2*) improved throughput through the presses (depending on the type of process, batch or continuous); *3*) improved lees settling and clarification rates, and *4*) reduced must viscosity for improved downstream processing. The processing of "difficult to press" varieties has often required the use of some form of a press aid, such as the back addition of stems to the press. Macerating enzymes decrease the "sliminess" associated with these varieties by breaking down the pectins and hemicelluloses responsible for this characteristic, resulting in increased plant throughput or capacity by 10% to 30%. Even with "normal" grapes, plant capacity can be increased from 5% to 30% depending on whether the process is a continuous (screw press, 5% to 15%), or a batch-type (basket press, 10% to 30%) process. The reduced must viscosity offers a number of downstream processing benefits. The quicker lees settling and clarification rates allow for faster racking of the lees by 10% to 100%. The decreased level of insoluble solids in the must benefits the processor by decreasing the need for fining agents and filtration aids by 10% to 100%. (Genencor International Inc., unpublished data, *21*)

Quality Improvements. Macerating enzymes have also been shown to improve the quality of the final product. The types of quality improvements observed are: *1*) improved color extraction in red grape varieties; *2*) increased aroma and flavor extraction in both red and white grape varieties; and *3*) improved body and mouthfeel of the finished wine.

Color extraction for both wine and grape concentrate is very important when one considers the fact that most products are held for extended periods of time and the determination of their overall quality is at least partially based on color. In the grape concentrate industry, color standards have already been established to an extent that the overall return to the processor is based on the amount of color present in the concentrate. Therefore, color extraction is extremely important in the final product. Depending on the stability of this color at the time of sale, the processor may be required to supplement the product to meet the accepted color standard for the concentrate being sold. It is also well known that during the aging process of wine or during the storage of grape concentrate, color will decrease over time, eventually stabilizing at a level significantly lower than that at the time of crush.

Macerating enzymes have been shown to increase the amount of color released from the skins by 20% to 150%, depending on the skin contact time, processing temperature, enzyme dosage and grape variety (Genencor International, Inc., unpublished data). Figures 4 and 5 demonstrate the color stability results for a 1990 vintage red wine (*22*). Figure 4 clearly shows the benefit of the use of macerating enzymes for this particular red varietal. The use of macerating enzymes improved the amount of color extracted as determined by total anthocyanin content by 10% over the control. Both the control and the enzyme-treated sample lost a significant amount of color after nine months (48% and 38%, respectively). The net color loss between the control and the enzyme treated samples had increased from a delta of 10% to a delta of 31%. The anthocyanin content for the enzyme treated sample stabilized at 170 to 180 ppm after nine months. The control or untreated sample continued to lose anthocyanin content after one year. The actual color loss, as measured by total anthocyanin content, was determined to be 52%. Figure 5 shows additional color stability data for this 1990 vintage red wine. The time-dependent stability of the hue, measured as the ratio of the absorbance at 520 nm to the absorbance at 420 nm, is shown here. At the six-month time point, the hue of the control demonstrated an 8% improvement over the hue of the enzyme treated sample, 1.63 cf. 1.5. After one year, the hue of the control had decreased to a level lower than that of the sample treated with macerating enzyme. Based on the data presented, it is anticipated that the control will continue to lose hue due to increased browning of the wine, whereas the hue of the enzyme treated sample should remain essentially unchanged. This color trend has also been observed in the grape concentrate industry (*23*).

Macerating enzymes have also shown improvements in the flavor and body of red wines as determined by a 10% to 30% increase in the level of tannins and proanthocyanidins

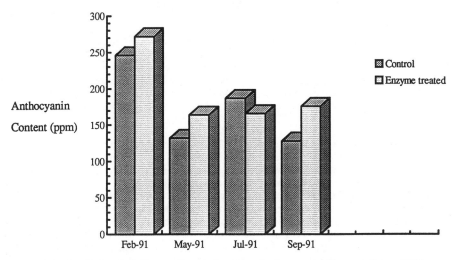

Anthocyanin
Content (ppm)

Figure 4: Color Stability as Determined by Anthocyanin Content for a 1990
Vintage Red Wine

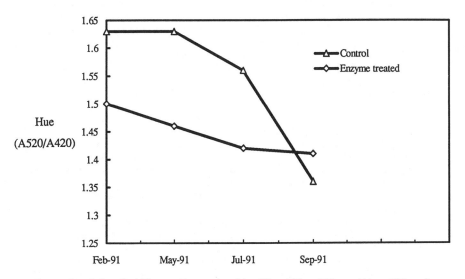

Hue
(A520/A420)

Figure 5: Color Stability as Determined by Hue (Abs. 520 nm/Abs. 420 nm)
for a 1990 Vintage Red Wine

present (*22*). Sensory evaluation supports the use of macerating enzymes to improve the overall quality of both red and white wine. (*22-23*). This has allowed the winemaker to obtain a better price for the finished product.

Commercial Enzyme Preparations: Proper Selection and Usage

A major hurdle faced by all commercial enzyme suppliers is the proper education of their customers or end users concerning proper selection and use their enzyme products. In order to do this successfully one must answer a number of important questions regarding the customer's process, and the desired outcome or the expected result from enzyme usage. The following questions should be addressed by the processor: *1*) What type of wine is being produced? *2*) What are the processing conditions being used? *3*) What are the economic considerations? *4*) What is the desired outcome? Each of these questions will be addressed in turn.

What Type of Wine Is Being Produced?   This question actually includes a number of more detailed questions regarding the outcome desired for the finished product. Is the end user processing red or white grapes to produce wine or grape concentrate? Is the end user trying to make full-bodied wines? Are they interested in improving the amount of color extracted from the skins? In order to accomplish all of these requirements in the process, one would recommend the use of a macerating enzyme which contains the necessary cellulolytic, hemicellulolytic and pectolytic activities. Most commercial enzyme preparations contain only pectolytic activity, which necessitates that these other side activities, cellulase and hemicellulase, be supplemented. If the processor is interested in reduced browning in the final product, commercially available laccase formulations from either Novo or Gist Brocades could find application (*24*). As stated in the previous section, macerating enzyme products have also resulted in decreased browning during processing and the aging process (*22*). Are they interested in flavor modification or enhancement in white wines? A number of commercial enzyme preparations are being touted for this application by the various enzyme suppliers. In particular these flavor enhancing/modifying formulations contain high levels of beta-glucosidase, alpha-arabinosidase, alpha-rhamnosidase, and alpha-apiosidase side activities. These flavor enhancing products will be discussed further in the section on future trends in the industry.

What Are the Processing Conditions Being Used?   Enzymes are catalytic entities which are affected by a number of processing parameters. In particular, time, temperature and pH have the greatest effects on enzyme efficacy. Enzyme efficacy can be directly translated into enzyme dosage requirements and, ultimately, into the end cost to the processor. In order to understand enzyme efficacy, one must have an understanding of enzyme pH and temperature profiles. Figures 6 and 7 depict typical pH and temperature profiles for a commercial macerating enzyme. In figure 6 the % relative activity or enzyme efficacy is measured against pH. All data has been normalized to pH 2.8 as 100% relative activity. This represents the lower end of the pH range encountered in winemaking. As the pH increases into the typical processing pH range for grapes (2.8 to 3.6), enzyme efficacy also increases. At the pH optimum for this particular macerating enzyme — pH 4.5 — the enzyme efficacy is represented as 200%. A similar trend is observed in figure 7. As the processing temperature decreases from the optimum of 50° C to 60° C to the typical processing range for grapes (red grape varieties — 25° C to 37° C and white grape varieties — 10° C to 15° C), the enzyme efficacy also decreases.

What does this mean to the end user? Decreases in both processing pH and temperature require increased enzyme dosages resulting in increased enzyme costs. Necessarily the opposite effect also holds. As the processing pH and temperature increase, the enzyme dosage required to obtain the desired effect will decrease. Adjustments of pH and temperature are not practical or desired solutions to the processor. However, the one parameter which can counterbalance the effects of both pH and temperature is the enzyme/berry contact time.

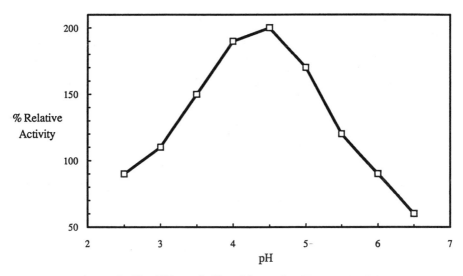

Figure 6: The Effect of pH on Macerating Enzyme Activity

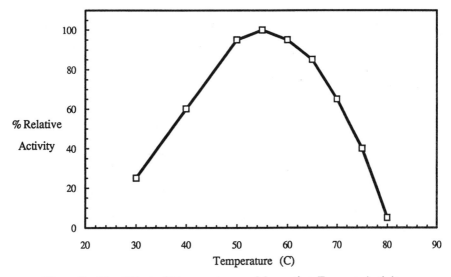

Figure 7: The Effect of Temperature on Macerating Enzyme Activity

Figure 8 gives some recommended processing conditions for the use of macerating enzymes showing the interaction of process temperature and skin contact time and their effects on enzyme dosage. In general, longer skin contact times and higher temperatures result in decreased enzyme consumption.

What Are the Economic Considerations?   This topic is extremely important to all grape processors and winemakers. Whether or not the use of enzymes should be considered is determined by current production volumes and grape costs. Capital expansions to increase production capacity can be foregone with the use of macerating enzymes. As reported earlier, macerating enzymes have been demonstrated to increase press throughput for both continuous (5% to 15% increased throughput) and batch-type (10% to 30% increased throughput) processes using both "normal" and "difficult to press" grape varieties.

In regard to grape costs, the use of macerating enzymes will be of greater importance for higher priced varieties ($1500-$2000 per ton) than for lower priced varieties ($100-$400 per ton). This is due to the fact that the processor wants to recover as much juice or must from the grapes in order to get the best return on the investment. In addition to the grape costs, the processor must also consider the enzyme costs per ton of grapes being processed. In order for the processor to evaluate an enzyme and to finally select between the different competitors' products one of two approaches can be used: 1) Evaluate the different enzyme products on a dosage basis or 2) on a cost basis. Approach 1) does not take into account the enzyme cost and may result in higher end user costs. The most relevant approach is 2), as this directly evaluates enzyme efficacy based on the processor's actual enzyme costs. The evaluation results in a direct correlation between processor's enzyme costs per unit activity which is easily measured as increases in the desired processing effect, yield, color extraction, *et al*.

Future Trends Within the Industry

The previous discussion details the development of the biotechnological aspects of both grape and wine processing. The details associated with the natural progression from whole cell yeast fermentation to the cell-free enzyme extracts for both pectinases and macerating enzymes have been discussed. At this point it is appropriate to discuss the direction of biotechnological advances in wine and grape processing. The areas which are currently receiving a significant amount of attention by the industry are: 1) The use of enzymes for flavor enhancement and aroma modification; 2) development of wine proteases for the elimination of wine haze development in white wines; 3) immobilized enzyme systems; 4) ureases to degrade ethyl carbamate; 5) laccases to eliminate the browning associated with white wine production; and 6) tannases which can be used to eliminate tannins from the wine.

Enzymes for Flavor Enhancement and Aroma Modification.   The chemistry of wine flavor has received a considerable amount of attention. Specifically, the role of the mono- and disaccharide conjugates of terpenes, norisoprenoids, and shikimic acid derivatives on wine quality has been the subject of much research in recent years (5-8). The structure of these flavor precursors is shown in Figure 9. The results of these studies indicate that the conversion of these flavorless, aromaless, aglycone conjugates to the free terpenes, norisoprenoids, and shikimic acid derivatives are of great importance to wine quality, *i.e.*, flavor and aroma development (5).

The use of glycosidases for the flavor and aroma enhancement of wine was first reported in the 1950's as a result of anthocyanase usage (Rohm & Haas, Genencor International, Inc., unpublished data). Since little was known of the chemistry of wine flavor/aroma at that time, it was difficult to determine if glycosidases were directly responsible. It has only been recently that the reaction mechanisms responsible for this flavor development have been elucidated. It is currently well established that enzymes for flavor and aroma enhancement

Must Temperature (C)

		5 to 10	10 to 20	20 to 30
Skin Contact Time (h)	< 1	150 to 200 ppm	100 to 150 ppm	50 to 75 ppm
	2 to 4	100 ppm	75 ppm	50 ppm
	> 5	50 to 75 ppm	50 ppm	25 to 50 ppm

Figure 8: Recommended Enzyme Dosages for Varying Process Conditions

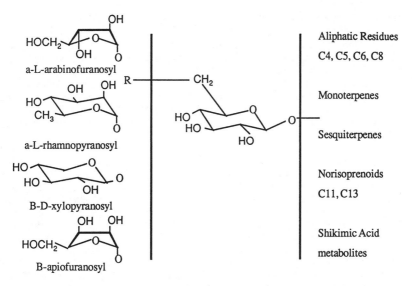

Figure 9: Important Flavor Precursors/Products in Wine Flavor Development (Reproduced with permission from ref. 10. Copyright 1990 The American Chemical Society)

must contain beta-glucosidase, alpha-arabinosidase, alpha-rhamnosidase, beta-xylanosidase, and beta-apiosidase activity (5-8, 10). In addition, these glycosidases should readily hydrolyze both the mono and diglycosides of primary, secondary, and tertiary alcohols, norisoprenoids, and shikimic acid derivatives. They must also be fully functional or uninhibited in the presence of high levels of both glucose and ethanol (Figures 10 and 11). They should also contain minimal side activities.

Flavor-enhancing enzymes can be added to either the must or the wine. The effect observed is dependent on the point of addition in the process. When the desired effect is achieved, bentonite treatment can be used to terminate the reactions. Obviously, a fair amount of testing is required by the winemaker in order to optimize the use of these enzymes.

Wine Proteases for the Treatment of Heat-Induced Wine Haze Development. Protein-aceous haze formation is a serious problem in white wine producing areas of the world. The proteins responsible for the instability have not been fully characterized. In 1959, Berg (25) presented a review on grape protein analysis which has further been augmented by both Heatherbell (26) and Williams (27). The haze-causing protein fraction has been determined to have a molecular weight range of 12,600 to 30,000 daltons. The classic method used to eliminate this proteinaceous haze is bentonite fining. There are a number of problems associated with bentonite fining: 1) large volume losses of wine from the lees, estimated at 5% to 10%, 2) waste disposal costs, and 3) overfining of wines resulting in the stripping of flavor and aroma characteristics. Other mechanical methods, such as ultrafiltration (28-29), have also proven successful at eliminating this proteinaceous haze. To date a significant amount of work (30-32) has also been carried out on wine proteases which can circumvent the use of bentonite fining and, to date, a solution has not been found.

Immobilized Enzyme Systems. Work has been carried out to immobilize wine proteases (30) but have met with limited success. Immobilized yeast systems have been used quite successfully in the fermentation of both beer and to a lesser extent wine (Dr. Heikki Lommi, Cultor Ltd., Finnsugar Bioproducts, personal communication, 1991). Attempts have also been made to immobilize the flavor enhancing or modifying activities such as beta-glucosidase. These have also met with limited success (Dr. Yves Galante, Lamberti Spa.; Claudio Caldini and Professor Franco Bonomi, Universita di Milano, Italy; personal communication, 1992). However, this is an area which shows a great deal of promise, especially for sensitive systems where it is difficult to remove the enzyme when it has completed its function.

Ureases Used to Remove Ethyl Carbamate. In 1985, ethyl carbamate formation in for-tified wines became a public health concern in Canada (33). Limits for ethyl carbamate content have subsequently been imposed by Health and Welfare Canada: 30 parts per billion (ppb) for table wines, 100 ppb for fortified wines, 150 ppb for distilled spirits and 400 ppb for brandy made from stone fruit. A significant amount of research effort has been expended in both California and New York in an attempt to determine the source of the ethyl carbamate precursors — as well as enzymatic methods — to degrade this potential carcinogen.

The formation of ethyl carbamate has been reported (34-38) to be the result of either: 1) urea addition as a yeast nutrient supplement useful in "stuck" fermentations or 2) to free amino acids (citrulline and arginine) present during fermentation. In general, the ethyl carbamate content of red wines is typically higher than in white wines. This can be attributed to higher fermentation temperatures and longer skin contact times.

In regard to the use of an acid urease to reduce the ethyl carbamate content of wines, the results (39-42) indicate that the potential for ethyl carbamate formation can be reduced by removing the residual urea. This solution is limited by the level of detectability for urea in commercial table wines. The sensitivity of the current technology is around 1 ppm to 2 ppm using activated carbon for red and rose wines. This is well above the Canadian restrictions giving ethyl carbamate levels in the ppb range. A sensitivity of 50 ppb could

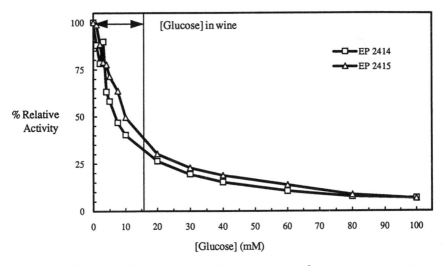

Figure 10: The Effect of Glucose Concentration on β-glucosidase Activity

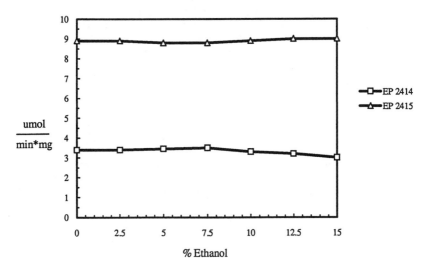

Figure 11: The Effect of Ethanol Concentration on β-glucosidase Activity

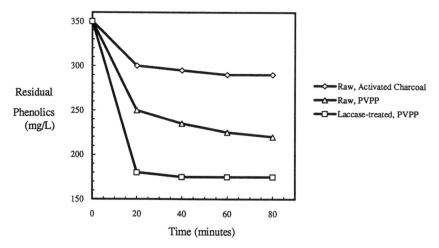

Figure 12: Phenolic Removal During Various "Active Filtrations", at Pilot-scale, on Raw and Laccase-treated Muscat Must (Reproduced with permission from ref. 46. Copyright 1989 Marcel Dekker, Inc.)

be achieved in white wines using a modification (*41*) of the method of Jansen (*44*) and utilizing the technology currently available.

There are a number of different schools of thought on the use of an acid urease for the removal of ethyl carbamate in wines. Yet, the consensus in the industry is that ethyl carbamate is currently not a problem in table wines. It has been suggested, however, that the use of diammonium phosphate (DAP) as a yeast nutrient supplement in place of urea, combined with proper wine storage, could help to circumvent this issue entirely. For this reason, the future for a urease in the industry does not appear very promising.

Laccases to Eliminate the Browning Associated with White Wine Production. Recently, the use of laccase in winemaking has been studied in an effort to induce phenolic browning. In theory, this would allow for the removal, through fining and/or filtration, of the phenolic compounds most susceptible to oxidation. Since this would occur prior to bottling, it should help to reduce phenolic oxidation during aging and storage.

Laccase is similar to the endogenous polyphenol oxidase (PPO), yet the two enzymes have quite different structures, reaction mechanisms, and substrate specificities (*11-12, 45*). In addition, laccase is significantly more stable than PPO. Recently, Cantarelli (*46*) reported results obtained using laccase in combination with PVPP, versus untreated must using other "active filtration" aids, charcoal or PVPP, on Muscat must. Figure 12 depicts the residual phenolics content versus time using various must treatments. The results clearly indicate that the use of laccase in combination with an active filtration aid, PVPP, significantly reduces the levels of phenolics in must. Once again, the results have been demonstrated in an academic setting, yet there are no indications of its use in commercial practice.

Tannases Which Can Be Used to Decrease the Tannin Content in Wine. In principle, tannases can be used to treat wines and remove tannin. As noted earlier, proanthocyanidins are another class of compounds often referred to as tannins. They can also affect wine quality due to their involvement in browning reactions and interactions with anthocyanins. There are no indications of the use of a tannase in commercial practice.

Summary

The previous discussion demonstrates some of the uses of biotechnology in grape processing and winemaking. This review is superficial in that not all topics are discussed fully in detail.

The level of sophistication or awareness of the processor continues to increase. A direct result of this is that the processor will be able to request a specific product to meet particular processing needs. As biotechnology progresses and some of the more common techniques available to the molecular biologist, such as genetic or protein engineering, become socially acceptable, this area will also continue to advance. Innovation through biotechnology will continue to meet the needs of the industry in the future.

LITERATURE CITED

1. Amerine, M.A.; Joslyn, M.A. *Table Wines: The Technology of Their Production;* University of California Press: Berkeley, CA, 1970.
2. Amerine, M.A.; Berg, H.W.; Kunkee, R.E.; Ough, C.S.; Singleton, V.L.; Webb, A.D. *The Technology of Wine Making;* AVI Publishing Co.: Wesport, CT, 1980.
3. Gallander, J.F. In *Chemistry of Winemaking;* Webb, A.D., Ed.; American Chemical Society: Washington, D.C., 1974; pp 11-49.
4. Brillouet, J.M.; Bosso, C.; Moutounet, M. *Am. J. Enol. Vitic.,* **1990**, *41(1),* pp 29-36.
5. Strauss, C.R.; Wilson, B.; Gooley, P.R.; Williams, P.J. In *Biogeneration of Aromas;* Parliment, T.H.; Croteau, R., Eds.; ACS Symposium Series No. 317, American Chemical Society: Washington, D.C., 1986; pp 222-242.
6. Williams, P.J.; Sefton, M.A.; Wilson, B. In *Flavor Chemistry Trends and Developments;* Teranishi, R.; Buttery, R.G.; Shahidi, F., Eds.; ACS Symposium No. 388, American Chemical Society: Washington, D.C., 1989; pp 35-48.
7. Versini, G.; Dalla Serra, A.; Dell'Eva, M.; Scienza, A.; Rapp, A. In *Bioflavour '87;* Schreier, P., Ed.; de Gruyter: Berlin, 1988; pp 161-170.
8. Gunata, Y.Z.; Bayonove, C.L.; Baumes, R.L.; Cordonnier, R.E. *J. Chromatogr.* **1985**, *331,* pp 83-90.
9. Litchev, V., *Am. J. Enol. Vitic.,* **1989**, *40(1),* pp 31-35.
10. Williams, P.J. *Hydrolytic Flavor Release from Non-volatile Precursors;* ACS Short Course, 200th National Meeting of the American Chemical Society, Washington, D.C., 1990.
11. Ribereau-Gayon, P. In *Chemistry of Winemaking;* Webb, A.D., Ed., American Chemical Society: Washington, D.C., 1974; pp 50-87.
12. Zent, J.B. *Controlling Enzymatic Browning in Foods Through the Use of Common Food Additives;* Master of Science Dissertation, Arizona State University, 1985.
13. Ussiglio-Tomasset, L. *Ann. Sper. Agrar. (Rome),* **1959**, *13,* pp 375-404.
14. Pilnik, W.; Voragen, A.G.J. In *The Biochemistry of Fruit and their Products;* Hulme, A.C., Ed.; Academic Press: New York and London, 1970, Vol. 1; pp 63-87.
15. Amerine, M.A.; Berg, H.W.; Cruess, W.V., *Technology of Wine Making;* 3rd ed.; AVI Publishing Co.: Westport, CT, 1972.
16. Robertson, G.L.; *Am. J. Enol. Vitic.,* **1979**, *30,* pp 182-186.
17. Nilov, V.I.; Datunashvili, G.M.; Zudova, G.M.; Ezhov, V.N.; Anachekova, G.M. *Sadov. Vinograd. Vinodel. Moldav.,* **1975**, *30,* pp 25-27. Abstract in FSTA 8:1H16, 1976.
18. Ussiglio-Tomasset, L.; Stefano, R. *Riv. Viticolt. Enol.,* **1977**, *30,* pp 452-469. Abstract in FSTA 11:1H56, 1979.
19. Villettaz, J. U.S. Patent 4,439,455, issued March 27, 1984.

20.   Wucherpfennig, K.; Dietrich, H. *Lebensmitteltechnik*, **1983**, *15*, pp 246, 248, 250, 252, 253.
21.   Flores-Gaytán, J.H. *Ultrafiltration of Fruit Juice and Wine;* Doctor of Philosophy Dissertation, Oregon State University, 1987.
22.   Zent, J.B.; Inama, S. *Am. J. Enol. Vitic.* **1992**, *43(3)*, 311.
23.   Haight, K.G.; Gump, B.H. *Am. J. Enol. Vitic.* **1992**, *43(3)*, 305.
24.   Novo Ferment (Switzerland), Ltd., Neumatt, 4243 Dittingen. Gist-brocades Food Ingredients, Inc., King of Prussia, PA
25.   Berg. H.W. Presented at the Tenth Annual Meeting of the American Society of Enologists; Fresno, CA, June 26-27, **1959**; pp 130-134.
26.   Heatherbell, D.A.; Flores, J.H. *Proceedings from the Second International Cool Climate Viticulture and Oenology Symposium;* Auckland, New Zealand, **January 1988**; pp 243-246.
27.   Waters, E.J.; Wallace, W.; Williams, P.J. *Am. J. Enol. Vitic.*, **1991**, *42(2)*, pp 123-127.
28.   Cattaruzza, A.; Peri, C.; Rossi, M. *Am. J. Enol. Vitic.*, **1987**, *38(2)*, pp 139-142.
29.   Flores, J.H.; Heatherbell, D.A.; Hsu, J.C.; Watson, B.T. *Am. J. Enol. Vitic.*, **1988**, *39(2)*, pp 180-187.
30.   Bakalinsky, A.T.; Boulton, R. *Am. J. Enol. Vitic.*, **1985**, *36(1)*, pp 23-29.
31.   Modra, E.J.; Williams, P.J. *The Australian Grapegrower and Winemaker*, April **1988**, pp 42-46.
32.   Waters, E.J.; Wallace, W.; Williams, P.J. In *Proceedings from the Seventh Australian Wine Industry Technical Conference, 13-17 August 1989: Adelaide, S.A.;* Williams, P.J.; Davidson, D.M., Eds.; Australian Industrial Publishers: Adelaide, S.A., 1990; pp 186-191.
33.   *Practical Winery and Vineyard;* Sept/Oct 1990, pp 8-12. Edited version updated from *Wine East,* **Nov/Dec 1989.**
34.   Tegmo-Larsson, I.-M.; Spittler, T.D.; Rogriguez, S.B. *Am. J. Enol. Vitic.*, **1989**, *40(2)*, pp 106-108.
35.   Tegmo-Larsson, I.-M.; Henick-Kling, T. *Am. J. Enol. Vitic.*, **1990**, *41(3)*, pp 189-192.
36.   Tegmo-Larsson, I.-M.; Henick-Kling, T. *Am. J. Enol. Vitic.*, **1990**, *41(4)*, pp 269-272.
37.   Tegmo-Larsson, I.-M.; Spittler, T.D. *J. Food Science*, **1990**, *55(4)*, pp 1166-1169.
38.   Monteiro, F.F.; Bisson, L.F. *Am. J. Enol. Vitic.*, **1991**, *42(3)*, pp 199-208.
39.   Ough, C.S.; Trioli, G. *Am. J. Enol. Vitic.*, **1988**, *39(4)*, pp 303-307.
40.   Fujinawa, S.; Burns, G.; de la Teja, P. *Am. J. Enol. Vitic.*, **1990**, *41(4)*, pp 350-354.
41.   Fujinawa, S.; Todoroki, H.; Ohashi, N.; Toda, J.; Terasaki, M. *J. Food Science*, **1990**, *55(4)*, pp 1018-1022.
42.   Kodama, S.; Suzuki, T.; Fujinawa, S.; de la Teja, P.; Yotsuzuka, F., presented at the *43rd Annual Meeting of the American Society for Enology and Viticulture*, Reno, NV, **June 24-27, 1992.**
43.   Boehringer Mannheim GmbH. **1987**, Urea/ammonia UV-method No. 787.961.2.576 964(1).
44.   Jansen, H.; Frei, R.W.; Brinkman, U.A.Th. *Qual. Plant. Foods Hum. Nutr.*, **1985**, *90*, 97.
45.   Schwimmer, S. *Source Book of Food Enzymology;* AVI Publishing Co.: Westport, CT, 1981.
46.   Cantarelli, C.; Brenna, O.; Giovanelli, G.; Rossi, M. *Food Biotechnology*, **1989**, *3(2)*, pp 203-213.

RECEIVED March 5, 1993

Chapter 11

Ultrafiltration

A New Approach for Quality Improvement of Pressed Wine

A. J. Shrikhande and S. A. Kupina

Heublein Wines, 12667 Road 24, Madera, CA 93639

The pressed wines are an integral part of grape processing and vary in quality depending upon pressing practices. Pressed wines are inferior to free run wines due to their higher polyphenol content. The principal phenolic compounds responsible for the astringency and bitterness in pressed wines are procyanidins.
Ultrafiltration is a superior alternative to conventional fining treatments and anion exchange for quality improvement of pressed wines.
Ultrafiltration was found to be more selective in removing pressed character while retaining most of the wines original varietal characteristics.
Profiles of the phenolics were obtained on High Performance Liquid Chromatography (HPLC) which was necessary for optimizing the ultrafiltration process for pressed wines.

Ultrafiltration Process Technology was investigated for wines which were characterized as intolerably harsh and astringent with strong pressy character. The objective was to develop a suitable technology to reduce harshness from these wines while maintaining sufficient fruitiness to enable blending of these wines as standard white wines.

Many classical approaches, including gelatin fining and anion exchange, were also evaluated for quality improvement of these wines. Gelatin fining at a rate of 10-15 lb/1000 gallons was needed to decrease harshness. This treatment also reduced the fruity character of wines and resulted in an unmanageable increase in lees volume and bentonite requirements. Anion exchange technology as well as a number of polymeric adsorbents proved futile and resulted in complete loss of fruitiness in pressed wines.

0097–6156/93/0536–0197$06.50/0
© 1993 American Chemical Society

All of the approaches mentioned above fell short of expectations and necessitated a novel approach which would selectively remove harsh components without appreciably affecting the fruitiness of pressed wines.

This research was successful in the application of an Ultrafiltration Technology for selectively reducing harshness and astringency while substantially maintaining the fruity character of hard pressed white wines.

This report describes the detailed research efforts made towards the applications of the Ultrafiltration Membrane Technology for the quality improvement of wines.

Chemistry Of Pressed Wines

The pressed wines are an integral part of grape processing and vary in quality and quantity depending upon the handling and pressing practices. Pressed wines are generally inferior to free run juice wines due to their higher polyphenol content. It is also generally recognized that longer the contact time of grape juice with skins and seeds, the larger the concentration of phenolics that are extracted into the wine.

The principal phenolic compounds responsible for the astringency and bitterness in wines are classified in the general group called flavonoids. A flavonoid is any compound containing a carbon-15 three ring base structure.

Basic Flavonoid Structure

The modification of central ring defines the subclassification. These compounds are exclusively found in the skins, seeds and stems of grapes. Approximately one-third are found in the skins and two-thirds are found in the seeds (1).

The two major subclasses of flavonoid that cause bitterness are flavonols and flavan-3-ols. The flavonols have the following basic structure.

Flavonol

The typical compounds in this category known to be present in grape skins are rutin, quercetin, quercitrin, myricitrin and kaemferol.   These compounds are associated with the bitterness of wine and range in molecular weight from 300-600.

The other category called flavan-3-ol is perhaps more important in wines and also responsible for bitter sensations.

Flavan-3-ol

The flavan-3-ol monomeric compounds such as (+)-Catechin, (-)-Epicatechin, (+)-Gallo-catechin (-)-Epigallocatechin are only found in the skins and the seeds of grapes.   The molecular weight of these compound is in the vicinity of 300.

Perhaps the most important group of flavonoid compounds found in wines belong to a polymerized form of flavan-3-ol.   These compounds in enological research are referred to as procyanidins (*1*).   These compounds are abundant in skins and seeds and get extracted in grape juice depending upon the severity of processing conditions.   The degree of polymerization of these compounds varies from simple dimers, trimers to heptamers (*3-4*).   The following structures are examples of a dimer and a trimer.

Dimer

Trimer

The molecular weight of these compounds varies from 600 for the dimers to 2100 for a heptamer.

The total phenolics present in wines also include some compounds referred to as non-flavonoid. These phenolic compounds are commonly referred to as cinnamates.

Cinnamic Acid

The four major cinnamic acid derivatives have been reported in white wines and they are caffeoyl tartaric acid, p-coumaryl tartaric acid, caffeic acid and p-coumaric acid (5-6). It has been shown that cinnamates are largely confined to the free run juice. They, therefore, constitute a much higher proportion of the total phenols of white wines and appear to be a major class of phenols in white wines made without pomace extraction. In comparison to the flavonoid compounds, the flavor effects of cinnamates in white wine are evidently mild. The molecular weight range of cinnamates in white wine varies from approximately 160 to 400.

In wine tastings, the sensations of astringency and bitterness are frequently confused. Astringency is identified as a puckery tactile mouthfeel sensation while bitterness is a true taste sensation. Recently, Lea and Arnold (*4*) had defined these sensations more critically.

Astringency. Is believed to result from non-specific and somewhat irreversible hydrogen bonding between o-diphenolic groups and protein in the mouth, thereby causing the distinctive drying and puckering sensation which is difficult to remove and makes further taste assessment a problem.

The larger the procyanidin concentration, the greater its capacity for hydrogen bonding and more astringency will be perceived.

Bitterness. Is regarded as being due to an interaction between polar molecules and lipid portion of the taste papillae membrane and hence it is critically dependent on the relative lipid solubilities of the bitter materials.

In the case of procyanidins, only the smaller molecules (up to tetramers) would be sufficiently fat soluble to pass in the lipid membrane and interact suitably with taste receptors to produce the phenomenon of bitterness.

The following conclusions have been drawn by Lea and Arnold (*4*) about the relationship of astringency and bitterness associated with monomeric catechins and polymeric procyanidin compounds.

1.   Astringency is predominately associated with the procyanidins that have a degree of polymerization greater than four.
2.   Bitterness is associated with monomeric catechins and procyanidins with degree of polymerization up to tetramer.
3.   There is no one procyanidin which is exclusively bitter and another which is exclusively astringent.

Singleton and Noble (*2*) suggested that the balance of bitterness and astringency in wines was concentration dependent too, so that the perceived bitterness is masked by a greater perceived astringency as the total procyanidin content increases.

Definition Of Ultrafiltration

Ultrafiltration (UF) is a process where a semi-permeable membrane separates the components of the liquid/solute mixture according to their molecular size. In ordinary filtration, the process liquid flows perpendicular to the filter; in ultrafiltration the process liquid flows tangential to the membrane.

The basic principle of the ultrafiltration operations is illustrated in Figure 1.   The solution containing two solutes flows tangential to the membrane; one solute's molecular size is too small to be retained by the membrane and the other is of larger size allowing retention by the membrane. A hydrostatic pressure is applied to the upstream side of the supported membrane and the solvent containing the small molecule solute passes through the membrane while larger molecular solute is rejected or retained by the membrane.

A feature of ultrafiltration, perhaps unique among filtration processes, is the ability to operate with steady filtration fluxes in absence of an external

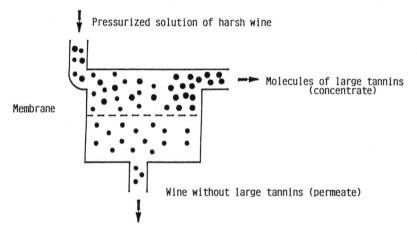

Figure 1.   Schematic diagram of membrane ultrafiltration process. (Reproduced with permission from ref. 9. Copyright 1977 Marcel Dekker, Inc.)

means for clearing the filter of accumulated solids. In ultrafiltration, the retained material always concentrates at the membrane-solution interface but is swept away by fluid dynamic forces.

The retained particle size is one characteristic distinguishing ultrafiltration from other filtration processes. Viewed on a spectrum of membrane separation processes, ultrafiltration is one of the membrane methods that can be used for molecular separations. In Figure 2, membrane size filtration is shown as a function of filtration flux. At the low flux end of the spectrum lie the commercial cellulose acetate reverse osmosis membranes with the capability of retaining sodium and chloride ions. Next come ultrafiltration membranes with the pores that span a range of 10^{-3} to 10^{-2} μm (10-100 Å) with filtration fluxes of about 0.5-10 gallons per square foot per day (GFD) per pound per square inch of driving pressure. Ultrafiltration membranes are commercially available for the molecular retention or separations in the range of 500 to 100,000 molecular weight cut offs. Microporous filters capable of virus and bacteria retention cover the size range of about .01-1.0 μm with fluxes of 10-1000 GFD. Finally, conventional filters for normal particulate materials are capable of filtering particles of 1 μm or larger with filtration fluxes above 1000 GFD.

Ultrafiltration Applications. The single largest application of UF has been in the processing of cheese whey. Ultrafiltration is used for the recovery of whey proteins as a by-product.

Ultrafiltration is also being successfully used for apple and pineapple juice clarification and is also being commercially used for enzyme, blood plasma, vaccine, hormone and variety of biomolecular separations.

Ultrafiltration Concept For Pressed Wines. Before considering UF for pressed wine it was relevant to study the molecular weight composition of pressed wine components (Table I) irrespective of percent composition of each fraction. The basic difference between the free run juice wine and pressed juice wine is due to higher concentrations of harsh and astringent phenolic compounds emanated from the skin and seeds in pressed juice. These compounds vary in molecular weight from 600-2000 and are the largest size molecular species in clarified wine. These compounds are far separated in wine from other phenolics ranging in molecular size between 200-500 and from sugars, acids, alcohol and aroma compounds which are in the vicinity of 200 molecular weight. In free run juice wines, the harsh and astringent compounds are present in low concentrations due to minimum contact with seeds and skins.

It was postulated that ultrafiltration with proper molecular membrane cut off in the vicinity of 1000 could selectively remove these larger harsh and astringent phenolic compounds without appreciably affecting the other basic wine components. With this basic premise of eliminating harsh and astringent phenolic compounds from pressed wine, the ultrafiltration selective membrane concept was explored and development proceeded into a feasible technology for wine processing.

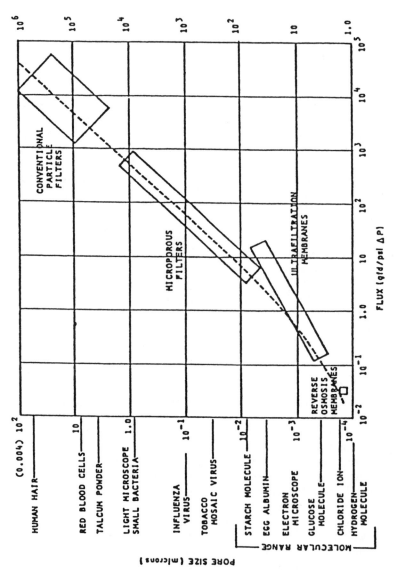

Figure 2. Pore size vs. flow rate for separation media. (Reproduced with permission from ref. 9. Copyright 1977 Marcel Dekker, Inc.)

Table I. Wine Composition as Related to Molecular Weight of Components

Wine Component	Molecular Weight
water	18
alcohol	46
reducing sugars	180
acids	150
aroma compounds	<300
simple phenols	<200
cinnamic acid derivatives	<400
catechins	300
simple flavonoids	500
procyanidins (dimers to heptamers)	600 - 2100
oxidized colored bodies	2000 - ?

Experimental Section

Hard Pressed Wine. Hard pressed wines, white and red, were evaluated for their potential improvement with ultrafiltration. The hard pressed wines tested for improvement was a mixture of dry white wine prepared from the hard pressed juice cut from the north coast plants. Approximately 15-20 gallons/ton of this juice from both the plants was normally mixed and fermented to produce this hard pressed dry white. This wine was extremely (intolerably) astringent and bitter due to abnormally high tannin content resulting from aggressive pressing with Coq presses.

For conducting a series of experiments, 1,000 gallons of this wine was shipped from the plants to the R&D Technical Center pilot plant and stored in the two 500 gallon double jacketed Mueller tanks. The wine was maintained at 35-40°F by cooling the tanks with chilled glycol.

Experiments were also conducted on 100 gallons of distilling grade red pressed wine obtained from a central valley winery.

Sulfur Dioxide. All the wines were periodically adjusted to 50 ppm free SO_2 by using a 6% solution of SO_2. This ensured prevention of any oxidation. These wines were also checked for dissolved oxygen periodically and sparged with nitrogen regularly to keep the dissolved oxygen below 2 ppm.

Protein Stabilization. All of the wine was protein stabilized by using 4 lb bentonite/1000 gallons, well racked and filtered using a plate and frame pilot plant assembly.

HPLC Profile of Phenolics. The qualitative profiles of phenolics by high performance liquid chromatography (HPLC) were conducted for control and ultrafiltered permeate samples of hard pressed wine. A noticeable broad band in the chromatographic phenolic profile particularly in the hard pressed wine (Figure 4.) was shown to be due to the combined presence of high molecular weight procyanidin compounds. Further resolution of this broad band was tried by adsorption of procyanidins on Sephadex LH-20 and successive elution and elimination of other phenolic compounds. This broad band was selectively eluted from LH-20 (Figure 5.) and was further subjected to HPLC separation.

Chromatography. A Water's High Performance Liquid Chromatograph was used. It consisted of two Model 6000A pumps, a Model 710B W1SP (autosampler), a Model 450 variable wavelength detector, a Model 720 system controller, and a Model 730 Data Module.

The column was a micro-Bondapak C-18 (10 micron) 3.9 cm ID x 30cm from Waters Associates.

Operating Conditions. Mobile Phase - All solvents were HPLC grade from J. T. Baker. Two solvent systems were used:

    A) acetic acid: water (2:98)

B)   acetic acid: acetonitrile: water (2:30:68)

The initial conditions were 0%B followed by a concave gradient curve 7 to 100%B in 35 minutes. 100%B was continued isocratically to 50 minutes. Curve 7 was used to return to initial conditions in 5 minutes, for the next injection.

flow rate	- 2 ml/min
U.V. detector wavelength	- 280 nm .04 AUFS
temperature	- ambient
injection volume	- 50 microliters
chart speed	- 0.5 cm/min
run time	- 55 minutes

Procedure.   A wine sample was filtered through a 0.45 micron filter into a sample vial. Sample vials were placed in the carousel and automatically injected with the WISP autosampler.

Procyanidin Quantitation.   The broad band of procyanidins obtained by HPLC analysis was difficult to quantitate. Further research is required to develop methods in resolving individual polymers into symmetric peaks by HPLC. However, a colorimetric reaction between vanillin and procyanidins (7) was found to be satisfactory and applied to our wine analyses. The method was quantitative and the procyanidin concentration is expressed as catechin in parts per million (mg/L).

Color.   During the ultrafiltration processing, the samples of original wine (control), concentrate samples and permeate samples were periodically monitored for 520 and 420 nm absorbances in a Gilford Spectrophotometer with a 1 cm path length cell. The control and concentrate wines were filtered through a Millipore 0.45 micron filter before color measurements. The permeate was used directly since it was always brilliantly clear.

Other Constituents.   The other constituents in wine such as alcohol, reducing sugars, titratable acidity, pH, total phenolics (Folin Ciocalteau) flavonoids and non-flavonoids were analyzed by well known standard analytical procedures.

Vanillin Assay for Procyanidins

Reagents.   Hydrochloric Acid (conc.) (J. T. Baker)
Glacial Acetic Acid (HPLC grade) (J. T. Baker)

Vanillin (Sigma Chemical Company)
(+)-Catechin (Sigma Chemical Company)

Apparatus. Constant temperature water bath.
UV/Visible Spectrophotometer
(Coleman 124D, Perkin Elmer)
Vials (10ml) snap-on-caps (VWR)
A 1 ml Eppendorf pipette.
A 5 ml serological pipet.

Procedure. 1) Prepare a constant 30°C water bath.
2) Prepare catechin standards of 100, 500, 1000 ppm for calibration curve.
3) Prepare an 8% HCl (w/v) (18.21 ml of HCl/100 ml) solution in acetic acid solvent. Also a 1% (w/v) vanillin solution in acetic acid.
4) Mix equal volumes (1:1) as needed of the 8% HCl and 1% vanillin solutions. Resulting concentrations are 4% HCl and 0.5% vanillin.
5) Take a 10 ml snap-on-cap vial and add 1 ml of sample using the 1 ml Eppendorf pipette.
6) Add 5 ml of the mixed vanillin and HCl solution to sample vial and mix well.
7) Place sample vial in 30°F water bath for 25 minutes and read absorbance at 510 nm.
8) Use a blank, 1 ml distilled water with 5 ml of mixed solution to zero the spectrophotometer.

Note: Normally absorbance goes off scale for control hardpressed wines. A dilution of 1:2 will bring absorbance on scale.

Calculations. The equation of the line from the standard curve was used to determine procyanidin content expressed as catechin ppm.

$$\text{Absorbance} = (\text{slope})(\text{catechin ppm}) + \text{Y intercept}$$

$$\text{Catechin ppm} = \frac{\text{Absorbance - Y Intercept}}{\text{Slope}}$$

Note: To convert Catechin ppm to Gallic Acid Equivalents (GAE ppm) a multiplication factor of 0.88 was used.

Results and Discussion

Ultrafiltration of Hardpressed Wines. In the first phase of this research program, emphasis was placed on the quality improvement of hardpressed wines.

A representative wine selected for quality improvement was a dry white press. The wine was intolerably harsh and had desirable distinctive fruity character. The research objective was to reduce harshness while maintaining the desirable fruitiness of the wine. As already mentioned in the Chemistry Section of this report, the harshness of these wines may be directly related to concentration of the polymeric phenolic compounds normally referred to as procyanidins.

The preliminary research was carried out by using non-ionic adsorbents, gelatin fining and anion exchange resins to remove these harsh astringent phenolic compounds. Efforts with polymeric adsorbents proved futile and resulted in complete loss of fruitiness in the pressed wine. Gelatin fining at the rate of 10 to 15 pounds per 1000 gallons was needed to decrease astringency. This gelatin level resulted in decreased fruitiness with an unmanageable increase in lees volume, protein instability and bentonite requirements. The anion exchange technology also reduced the fruitiness of the wine excessively.

Verification of the New Approach. After realizing the deficiencies of the above processing methods, a new approach based on selective removal of harsh-astringent phenolic compounds employing ultrafiltration membranes was explored. A preliminary test was carried out with a polysulfone 1000 molecular weight (MW) cutoff membrane. This test validated the proposed concept that the ultrafiltration membrane technology could substantially reduce the undesirable harsh-astringent phenolic compounds from the pressed wines. This successful exploration of a new concept created a potential technology for wine treatment.

Ultrafiltration Process. After the verification of the concept, it was necessary to select proper UF systems and proper membranes to provide the following:

a)   The ability of the membranes to retain or reject undesirable harsh astringent compounds while permitting retention of maximum fruitiness in wines;

b)   The ability of the system and its membranes to produce steady high fluxes without appreciable fouling;

c)   The cleaning efficiency of the membrane in repeated use.

Various UF membranes were evaluated. Different pore size with molecular weight cut-off membranes, different membrane polymers (materials) were also evaluated.

The simplified ultrafiltration flow schematic used in testing the membranes is shown in Figure 3. The wine from feed tank was recirculated over the membrane tangentially and the permeate (soft wine) was recovered continuously. Since large polymeric tannins, brown pigments and residual proteins were essentially retained by the membranes, the retentate always had an increased concentration of these compounds.

The process continued by recirculation until most of the wine was recovered as permeate. The majority of the experiments in this study were carried out until 85-95% of the original feed volume was recovered as permeate.

Comparative Membrane Performance in Rejection of Harsh Phenolic Compounds. Since excessive harshness was a limiting quality attribute of hardpressed wines, various membranes were evaluated for their ability to reduce these harsh compounds in the wines.

A special quantitative method based on vanillin colorimetric test was developed which measured the total flavon-3-ol moities consisting of both monomeric catechins and polymeric catechins (procyanidins). It was also shown that hardpressed wines had considerable quantity of procyanidins with a molecular weight of more than 1200. This confirmation came from a HPLC qualitative profile of phenolics from hardpressed wines. The HPLC qualitative profile of hardpressed dry white can be seen in Figure 4. The shaded area represents the area occupied by procyanidins. The intact isolation of this area by preparative Sephadex LH-20 column chromatography further substantiated the occurrence of these compounds in hard pressed wines (Figure 5).

Different UF membranes were evaluated. Polysulfone hollow fiber membranes, with 1000, 2000 and 10,000 MW cutoff were tested. It was interesting to note that rejectivity of the procyanidins was similar for 1000, 2000 and 10,000 MW membranes. It was observed that the 65 to 70% procyanidin rejection by these membranes made the wines smooth and low in bitterness. However, other considerations such as extensive loss of fruitiness and low flux rates limited their commercial exploitation.

Cellulosic UF membranes were also evaluated. A 10,000 MW cutoff rejected approximately 64-69%, of the procyanidins while the 30,000 MW cutoff was less effective. On the basis of optimum procyanidin rejection, substantially more fruitiness retention and high flux rates, cellulosic membranes provided better performance.

A HPLC qualitative profile for an ultrafiltered hardpressed wine using a cellulosic membrane is presented in Figure 6. It is interesting to note that the zone of procyanidins (harsh phenolics) represented by the shaded area in hard press control wine is considerably reduced by this 10,000 MW membrane without appreciably affecting the other simple phenolic peaks before, over and after the shaded area in Figure 6. This indicated that procyanidin molecules with a degree of polymerization probably greater than 4 are effectively retained and rejected by the membrane (4). Processing the same wine with 30,000 MW cutoff membrane did not reduce the shaded area to the extent as the 10,000 MW cutoff membrane.

The above experiment demonstrated that a 10,000 MW membrane behaves very similarly to the tighter 1000 MW membrane. However, greater than 10,000 MW cutoff membranes are less effective.

It is interesting to note that although the actual molecular weight of the procyanidins in white wine never exceeds 2100, they are still retained by 10,000 MW cutoff membranes. This is probably due to two reasons: (a) The intermolecular association of procyanidins due to weak hydrogen bonding creates much larger molecular size; and (b) The branched structure of these compounds apparently inhibits the passage through the membrane pores (see structure of procyanidin trimer).

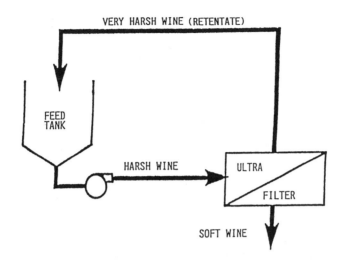

Figure 3. Simplified ultrafiltration flow schematic.

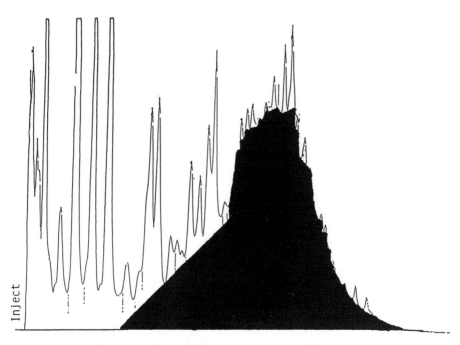

Figure 4. Control, hard pressed dry white wine.

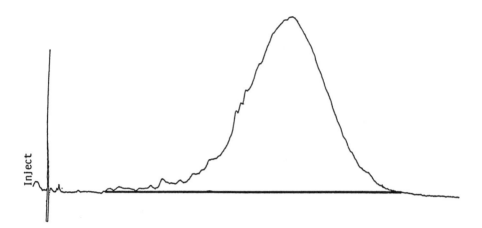

Figure 5. Isolated procyanidins (>4 units) on Sephadex LX-20 from hard pressed dry white wine.

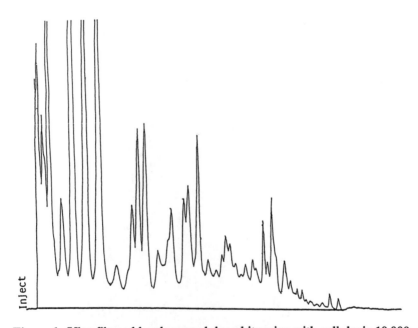

Figure 6. Ultrafiltered hard pressed dry white wine with cellulosic 10,000 MW membrane.

Rejection of Color Molecules. As can be observed in Figure 7 the color of the permeate slightly increased as the recoveries progressed to 90-95%. However, a larger part of the color (A_{420}) was rejected by the membranes with a steady increase initially followed by a steep increase as the retentate developed high concentration factors. At 95% recovery the retentate developed a concentration factor of 20.

The color improvement of ultrafiltered wine in contrast to the brown appearance of the control wine and the extremely dark brown appearance of rejected retentate is evident in Figure 8.

Sensory Evaluation Tests for Selection of Proper Membrane. The sensory evaluation tests were regularly conducted to assess the performance of various membranes on the total overall quality of the ultrafiltered wines.

The earliest tests were comparisons of ultrafiltered wines treated with polysulfone membranes against laboratory anion exchanged hardpressed dry white wine. The ultrafiltered wine was judged superior to anion exchanged wine. However, the wine lost too much desirable fruitiness. Experiments with the polysulfone 10,000 MW cutoff indicated that these membranes reduced astringency and bitterness but were far from satisfactory in fruitiness retention.

The cellulosic 10,000 MW cutoff membranes were rated the best for control of astringency and bitterness while retaining desirable fruitiness in ultrafiltered wines. Further taste panel studies indicated preference for the cellulosic membranes over the polysulfone 10,000 MW cutoff membranes.

The taste panel also compared the cellulosic 10,000 MW cutoff treated dry white pressed wine against standard dry white wine and found that the ultrafiltered pressed wine was sufficiently improved to approach dry white wine quality.

Comparative Flux Performance. The processing performance of any UF system is dependent on two important parameters:

(a)   The ability of the system and its membranes to produce steady high fluxes (throughput); and
(b)   The ability of the membrane to retain or reject undesirable molecules from a fluid stream.

These parameters determine the economics of a UF system and should be considered important criteria when choosing a system.

The polysulfone membranes with increasing porosity from 1000 to 10,000 MW cutoff showed only a marginal increase in flux from 0.93 to 3.66 GFD. The cellulosic 10,000 MW cutoff membranes produced the highest flux of 17.6 GFD.

The reasons for low fluxes with polysulfone membranes may be attributed to instantaneous adsorption of wine phenolics resulting in a secondary film (gel layer) formation on the membrane surface which acts as a barrier for efficient flow. The reasons for the high flux with cellulosic membrane appears to be due to the general inertness or neutral nature of the cellulosic membranes towards wine phenolics, with little or no gel layer formation.

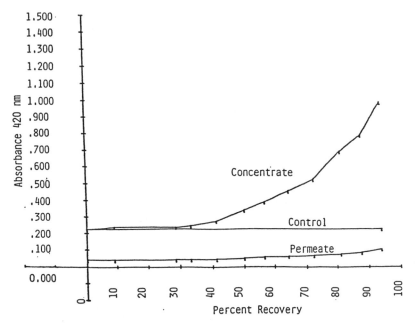

Figure 7. Performance of cellulosic 10,000 MW membrane in color removal from hard pressed dry white wine.

Figure 8. Visible appearance of (left) control hard pressed wine, (middle) ultrafiltered permeate, (right) retentate or rejected wine.

The cellulosic 10,000 MW cutoff membrane produces steady fluxes until about 90% recovery followed by a slight declining trend at higher recoveries. This decline in flux is probably associated with the increase in the viscosity of the wine due to excessive accumulation of retained phenolics and colored bodies.

Ultrafiltration of Red Pressed Wine. In red wine processing, the crushed grapes with skins and seeds are fermented and the free run juice wine is separated while the remaining skins and seeds are pressed to recover additional wine. This pressed wine becomes excessively harsh and is normally diverted to a distilling material which is eventually recovered as grape spirits. However, this diversion results in economic loss since a by-product of lower value is recovered in place of red wine.

Distilling grade red press wine was ultrafiltered with 10,000 MW cutoff cellulose membrane to 99.5 gallons of permeate (product) and 0.5 gallons of retentate portion. The permeate portion was essentially devoid of harshness while retaining other characteristics of original wine. Table II displays the effects of the ultrafiltration by comparing ultrafiltered red pressed wine and the retentate with a red pressed wine control. From Table II it is evident that the polyphenol content of original wine decreased from 3746 to 2851 in the ultrafiltered red pressed wine. It is apparent that the polyphenols (mainly procyanidins) are responsible for harshness. The color intensity in the ultrafiltered wine decreased from an original value of 1.96 to 1.43 which most probably accounts for those red pigments which are intimately associated with harsh polyphenols. These qualities together with practically little change in alcohol, acid, volatile acid and SO_2 reflect the specificity of the ultrafiltration membranes for harshness reduction from red wines. The ultrafiltered red pressed wine was improved in quality from distillation grade wine to standard red wine blend.

Effects of Ultrafiltration on the Essential Wine Components. The effects of different membranes on the total phenolics, flavonoids and non-flavonoids in the hardpressed wine, can be observed in Table III.

As expected, and consistent with the chemistry of tannins, only the flavonoid group of compounds and more specifically procyanidins were significantly reduced while little or no change was found in the non-flavonoid phenolics (Table III). Also, the retentate portion of ultrafiltered wines showed a greater increase in flavonoid concentration. This increase was dependent upon the percent recovery of permeate and the degree of retentate concentration in each experiment. Alcohol losses were insignificant in wines treated with cellulosic 10,000 MW cutoff membranes. This minimal loss of alcohol in certain experiments may be attributed to the continuous exposure of wine in the recirculation loop up to 60°F. The losses can be further minimized in a commercial system by maintaining a rigid temperature control at 45°F.

A slight decrease in free and total SO_2 was also observed in ultrafiltered wines and again can be minimized by maintaining a proper temperature control. The pH of the ultrafiltered wines remained essentially constant.

The color intensity of the hardpressed wine was considerably reduced by

Table II. Effects Of Ultrafiltration With Cellulosic 10,000 Molecular Weight Cutoff Membrane On Polyphenols And Other Constituents Of Red Pressed Wine

	WINE TREATMENT		
	Red Press (Control)	Ultrafiltered Red Press (Product)	Retentate, Rejected Portion of Wine
polyphenols (ppm)	3746	2851	26,966
intensity	1.96	1.429	1.306*
hue	1.57	1.58	1.59
alcohol % by volume	11.07	10.94	10.91
acidity gm/100 ml	.660	.635	1.01
SO_2	76	54	64
taste	very harsh, fruity	not harsh, fruity	excessively harsh

*Retentate diluted 10:1 with pH 3.5 buffer for color measurement

Table III. Effects of Ultrafiltration Membranes on Polyphenols of Hardpressed White Wine

Type of Membrane	Type of Sample	Total Phenolics (ppm)	Non-Flavanoids (ppm)	Flavanoids (ppm)	Procyanidins in GAE (ppm)
Polysulfone 1000 MW	Control	1330	235	1095	1355
Polysulfone 1000 MW	Permeate	739	345	394	491
Polysulfone 1000 MW	Retentate	8820	150	8670	
Polysulfone 2000 MW	Control	1303	188	1115	1355
Polysulfone 2000 MW	Permeate	658	243	415	408
Polysulfone 2000 MW	Retentate	5220	105	5115	
Polysulfone 10,000 MW	Permeate	478	299	179	512
Polysulfone 10,000 MW	Retentate	1529	217	1312	
Polysulfone 10,000 MW					
Cellulosic 10,000 MW	Permeate	559	341	218	536
Cellulosic 10,000 MW	Retentate	5454	98	5356	
Cellulosic 30,000 MW	Permeate	826			

ultrafiltration process providing distinct pale yellow color. The color intensity of the retentate portions always increased due to the rejection of the brown colored molecules by the ultrafiltration membranes.

The total acidity was marginally decreased in permeate samples and is probably associated with the phenolics reduction. All phenolic molecules are weak acids and confer some acidity to the wines. A decrease in acidity of permeate portion of wine and simultaneous rise in acidity of concentrate reinforces this theory that phenolics do confer acidity to the wines.

The reducing sugars were also marginally decreased in permeate samples with simultaneous increase in concentrate samples. This indicates some interference of phenolics in the reducing sugar determination.

Summary and conclusions.

Ultrafiltration appears to be a superior and viable alternative to anion exchange for upgrading the hard pressed white wines from distilling material to standard white wine quality.

Ultrafiltration is more selective in removing astringency from both white and red wines while retaining more fruit than conventional methods.

The 10,000 molecular weight cut off cellulosic membrane was found to be optimum for both white and red wine applications (8).

Acknowledgements.

This investigation was a coordinated effort of many individuals from the Technical Center. The contribution made by each scientist is gratefully acknowledged.

Joe Alioto coordinated the sensory evaluation tests which helped us in selecting a proper ultrafiltration system.

Literature Cited.

1. Arnold, R.A.; Noble, A.C.; Singleton, V.L. J. Ag. Food Chem. **1980**, 28, 675-678.
2. Singleton, V.L.; Noble, A.C. Advances Chem. **1976**, 26,47-70.
3. Lea, A.G.H.; Bridal, P.; Timberlake, C.F.; Singleton V.L. Am. J. Enol. Vitic. **1979**, 30, 289-300
4. Lea, A.G.H.; Arnold, G.M.J. Sci. Fd. Agric. **1978**, 29, 478-483.
5. Singleton, V.L.; Timberlake, C.F.; Lea, A.G.H. J. Sci. Fd. Agric. **1978**, 29, 403-410.
6. Okamura, S.; Watanabe, M. Agric. Biol. Chem. **1981**, 45, 2063-2070.
7. Butler, L.G.; Price, M.L.; Brotherton, J.E.J. Ag. Food Chem. **1982**, 30, 1087-1089.
8. Shrikhande, A.J. United States Patent, 4,834,998,1989.
9. Filtration Principles and Practices; Orr, C., Eds.; Part I; Marcel Dekker, Inc: New York, New York, 1977, Vol. 10; pp 477-479.

RECEIVED May 5, 1993

Chapter 12

Capture and Use of Volatile Flavor Constituents Emitted during Wine Fermentation

C. J. Muller, V. L. Wahlstrom, and K. C. Fugelsang

Department of Enology, Food Science, and Nutrition and Viticulture and Enology Research Center, California State University, Fresno, CA 93740–0089

Wine volatile flavor constituents originally present in the grape or produced during the fermentation process are often lost into the atmosphere during the vinification process due to their inherent volatility and to entrainment with evolved carbon dioxide. Many compounds thus lost have positive sensory attributes; others are often construed as detrimental to the aroma and flavor of the wine. Selective capture, separation, concentration and addition of these volatiles to the wines from whence they came can improve their quality.

Volatile compounds comprise a crucial constituency in determining the identity and complexity of wines. Many of these volatiles are present originally in the grape where they are produced by the myriad of biochemical interactions during development, growth and ripening. After harvest, other volatiles are generated by chemical and biochemical reactions arising from the inherent damage perpetrated on the grapes by the harvesting process. This is particularly true in the case of machine harvesting in which clusters are beaten off the vines and where considerable juicing occurs.

More often than not, there are delays from the time of harvest until actual processing prior to vinification beginning. During this time, further reactions occur. These reactions are accelerated by the relatively high ambient temperatures prevalent during harvest. Upon crushing, cell disruption allows enzymes and substrates to comingle freely.

0097–6156/93/0536–0219$06.00/0

In this environment, many new compounds are generated, a great portion of them volatile. It is at this time that skin lipids, for instance, might oxidize with the subsequent formation of short-chain alcohols and aldehydes. Surely, judicious use of enzyme inhibitors and antioxidants such as sulfur dioxide attenuate many of these changes.

Unless the grapes are of optimal quality, the winemaker must resort to high levels of sulfur dioxide to prevent undesirable changes. However, the current industry trend is for less and less use of sulfur dioxide at crush. Furthermore, by not using sulfur dioxide, microorganisms, mostly the native yeasts commonly found on grapes, might proliferate with the concomitant production of their own metabolites formed during incipient fermentation.

Thus, a very complex mixture of volatiles and their precursors already exist in the grapes prior to vinification. Some of these volatiles are intimately associated with the characteristic varietal aroma; others are associated with incipient processing changes, and finally, others are associated with definite chemical and microbiological spoilage. The extent to which these predominate is obviously primarily contingent upon the quality of the grapes at harvest, but also to the rapidity and care with which the harvesting, transporting, receiving and crushing operations are carried out and, above all, the ambient temperature.

Vinification as practiced in most commercial wineries is initiated by inoculating the must with either a known culture of actively growing wine yeast, often a freshly reconstituted wine active dry yeast (WADY), or by transferring actively fermenting must from another tank. Generally, white wines are fermented at about 12 deg C (55 deg F) whereas red wines are fermented at about 24-27 deg C (75-80 deg F). Obviously, both fermentation rate and thus evolution of volatiles, including carbon dioxide and ethanol are faster at the higher fermentation temperatures.

Fermentation Volatiles

The identity of the volatile constituents produced during fermentation has been the subject of much scrutiny (1,2,3,4,5,6). Researchers are continuously adding to this seemingly unending list. It is not the intent of this paper to dwell on each of the many compounds presently identified. Instead we intend to focus on the capture and utilization of those which emanate from fermentation tank vents while considering them as a group.

The amount and nature of the volatiles evolved during fermentation depends not only on fermentation temperature and grapes used and, as indicated above, on chemical and biochemical changes occurring prior to fermentation, but also on the type of yeast used (7). Yeast strains are known to differ markedly on their reducing ability. Some are capable of reducing sulfate to hydrogen sulfide (8,9). Yeast strains also differ in their ability to produce flavor constituents under normal conditions (10).

Other concurrent fermentations; e.g.: malo-lactic are definitely another source of volatiles emanating during the fermentation process. Many winemakers purposely induce a malo-lactic fermentation early, whereas others allow the yeast fermentation to either proceed or to be complete before inoculating with malo-lactic bacteria (11). Some of the volatiles emanating from such fermentations have definite positive sensory attributes.

Effect of Temperature

In general, volatile constituent production increases with increasing fermentation temperature. Also, so does the loss into the atmosphere of those constituents with the highest volatilities.

These include some very delicate aromas associated with the fruity and varietal characteristics of the grape as well as some of the compounds associated with the vinous character of the fermenting must.

Thus, white wines are traditionally fermented at lower temperatures in an effort to retain within the fermenting must as many of these volatiles as possible. Red wines on the other hand, are fermented at higher temperatures in an effort to extract as much color (and often tannins) as required for the type of wine being made.

Under these circumstances, it is customary to thoroughly mix solids and liquid ("must") by pumping from the bottom of the tank over the solids ("cap") which float on top of the fermenting liquid. This process is called "pumpover". It is done to extract color and, perhaps more importantly, to prevent the cap from drying. A dry cap often leading to the potential oxidation of ethanol and other metabolites to produce compounds such as acetic acid which, if present in large concentrations, are frankly objectionable. Surely, the process of pumping over itself allows volatile constituents to be lost into the atmosphere with great facility.

Yeast which are temperature stressed often produce uncommon volatiles, or higher amounts of undesirable volatiles such as fusel oils (12). Production of fusel oils is also enhanced when the yeast are stressed by lack of sugars, as occurs toward the end of fermentation.

Effect of Skin Contact

White wines are given very little skin contact by most winemakers. As indicated above, red wines on the other hand are given considerable skin contact during fermentation. Many winemakers prefer to separate the pomace well before the completion of fermentation whereas others allow the fermentation to reach completion prior to pressing.

Grape skins contain a complex mixture of carbohydrates, tannins and lipids with a small amount of protein. However, some of the most active grape enzymes are present in the skin. These include several oxygenases, polyphenol oxidases and lipoxygenases (13), all of which are capable of rapidly breaking down complex substrates to produce volatile constituents.

Thus, it is apparent that many potentially beneficial compounds are being produced which are simply lost into the atmosphere during the fermentation process. Were there ways to capture and reincorporate these constituents into the wine it is conceivably possible to increase the quality of such wine.

However, the process is not as simple as it seems inasmuch as constituents having positive attributes, are intimately comingled with many constituents detrimental to aroma and flavor. Any effort to enhance wine quality by reincorporating volatiles produced during fermentation must also be concerned with the separation of undesirable compounds or classes of compounds prior to incorporation.

Collection of Volatiles

The technology for capture of emission control volatiles has been implemented for a long time in the petroleum and chemical industries. These industries have been the focus of much scrutiny by the various government agencies entrusted with enforcing environmental quality.

However, only recently has the wine industry been subjected to similar scrutiny, and then only in California where the California Air Resources Board, and lately regional Air Quality Control Districts, have been investigating the extent to which wineries might contribute to atmospheric pollution (14).

It is in this regard that we, at California State University Fresno have been involved with studying emission control from winery fermentation tanks (15,16,17). As a direct result of these studies, we have in place emission control devices centered around charcoal adsorption traps. Such devices have been used to collect and capture fermentation emission volatiles for their identification and also for studies leading to wine quality enhancement (18).

Equipment

Pilot plant fermentation equipment used for these studies at CSU-Fresno, consists of four 1412-gallon stainless steel jacketed and insulated fermentors provided with stainless steel capture hoods as described elsewhere (17).

The capture hood is connected with 1-inch stainless steel square tubing to a foamover vessel. From here, the line goes first to a heat exchanger where the volatile stream is then cooled to about 4 deg C to remove as much moisture as possible, followed by a preheater. A rotary vane pump located downstream provides a slight reduced pressure onto the fermentation tank to remove volatiles. The same pump, on its discharge side, provides a slight pressure to direct the stream onto parallel-connected stainless steel vessels containing activated charcoal (see Figure 1).

Desorption

During normal operation, appropriate valving allows one of the charcoal adsorption vessels to be in the adsorption mode while the other is either being regenerated or is idle. Upon saturation of the charcoal with volatiles, regeneration and concomitant removal of the adsorbed volatiles is accomplished by directing clean, dry steam in counter-current fashion. The volatile-laden steam is thereafter condensed and collected in a stainless steel vessel (4 deg C). A dry ice stainless steel trap located downstream collects any uncondensed light volatiles.

Concentration of Volatiles

Volatiles captured as described above are then extracted 10 times with 50 mL aliquots of fluorotrichloromethane (Freon-11, Aldrich 24,499-1). The fractions are combined, dried over anhydrous sodium sulfate and the solvent evaporated in a rotary evaporator. The concentrated volatiles thus obtained are kept under nitrogen at -40 deg C until utilization.

We have not attempted a quantitative recovery of all volatiles emanating during fermentation. There is vast variability in the amount of volatiles produced in each and every fermentation tank. Subtle differences in temperature, degree of mixing and other factors provide for discrepancies in the fermentation rate and thus in the volatiles being produced and their time of emergence.

However, in our capture experiments we always attempted to trap as much of the volatile fraction as possible. It was apparent at the time collections were being made that some very ephemeral constituents were lost into the atmosphere regardless of all the precautions being taken to trap and collect them. Some of these constituents had floral connotations that undoubtedly would have made a very positive contribution to the aroma of the wine. Unfortunately, since they are so difficult to capture their identity and possible contribution is not known at present.

It is important to point out at this time that the whole process of capture, adsorption, desorption and concentration naturally selects some constituents at the expense of others. Therefore, the reconstituted volatile fraction only roughly resembles the actual constituency of the volatiles emanating from the fermentation tank vents

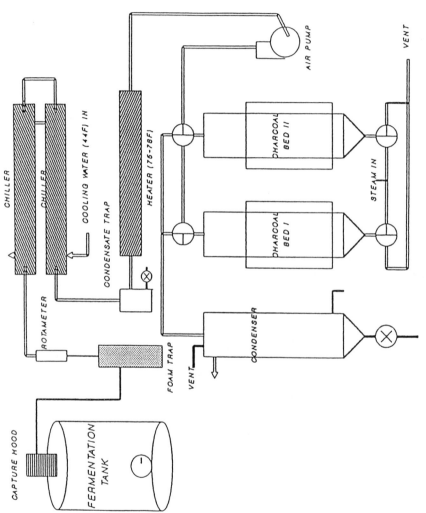

Figure 1. Volatile Collections System.

during the fermentation process. Furthermore, it is a well known fact that some constituents might exhibit positive attributes when in very dilute solution whereas in concentrated form they might be frankly detrimental to the overall aroma and bouquet sensation.

Also, it must be recognized that synergistic effects, both positive and negative, exist among compounds isolated in this fashion just like they exist in other environments. Thus the addition of the volatile fraction as a whole or in part to the wine poses a definite challenge inasmuch as the results are not always predictable.

Addition of Volatiles

In our initial work, the volatile fraction was collected, eluted from the charcoal, extracted from the aqueous phase with Freon-11, and the solvent evaporated as described above. No attempt was made at this time to separate the whole volatile fraction into subfractions by either boiling point or functionality. Instead, the volatiles thus obtained were added back to wines from which they came in various ratios and the resulting wines subjected to sensory evaluation by first, a trained panel, and then a "consumer" panel. The purpose here was to ascertain if such additions indeed improved the quality of the wines as expected. Both panels were requested to evaluate both intensity of aroma/bouquet and preference for each sample. Wine to which no volatiles had been added served as control.

Results

Results are shown graphically in Figures 2 and 3 for a white wine, and 4 and 5 for a red wine. Both wines were made from CSUF grapes and fermented dry. Data obtained therewith showed that it is indeed possible to achieve a modest improvement in the aroma/bouquet of both white and red wines (18).

However, it was also apparent that such improvement was contingent upon the ratio of addition of volatile fraction to wine. In the case of white wine for instance, addition of volatiles to wine in a 1:1 ratio, or 1X addition, resulted in wines with a lesser score both in terms of intensity and of preference than wines in which the addition was carried out at higher ratios. In each case the first figure in the ratio indicates the volume of fermenting must from which volatiles were captured to be added back; the second figure indicates the volume of wine to which those volatiles were added.

No significant increase (at $p < 0.05$) in aroma/bouquet intensities were found in white wines whose volatiles had been added in a 2:1 ratio (a 2X addition). However, preference tests on the aroma and bouquet characteristics

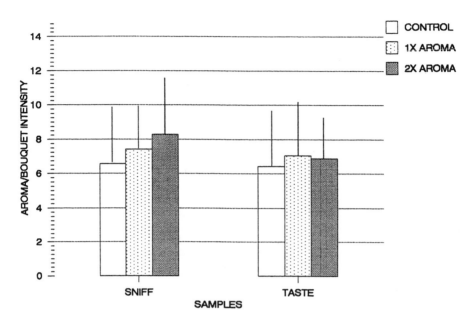

Figure 2. Aroma/Bouquet of White Wine (N=15x2).

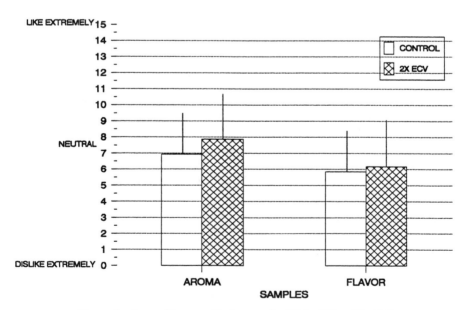

Figure 3. Aroma/Bouquet Preference for White Wine (N=106).

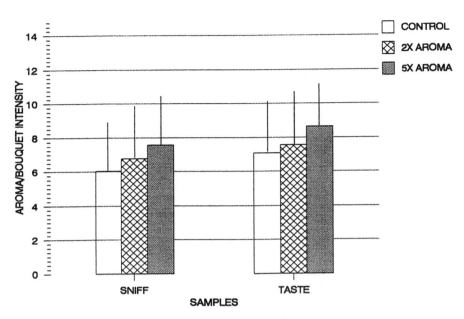

Figure 4. Aroma/Bouquet of Red Wine (N=15x2).

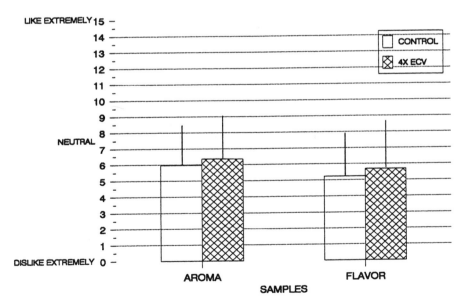

Figure 5. Aroma/Bouquet Preference for Red Wine (N=102).

were significant at p<0.05 with respect to the control
when the volatiles were added at higher ratios.

Results for the red wine were similar. In the case of
the red wine however, it was found that addition of
volatiles in a 4:1 ratio of volatiles to wine was
necessary to significantly enhance (at p<0.05) the aroma
and bouquet intensity of such wine. The same level of
addition also showed significant improvement of preference
characteristics (p<0.05).

However, such enhancement did not necessarily
translate into a frank improvement of the overall quality
of the wine. A typical comment of some members of the
trained panel was: "I perceive increased complexity and
stronger aroma and bouquet in these wines; however, I do
not always like the aroma and bouquet". Similar comments
were made about the overall taste of both the white and
the red wines. Such response from the trained panel was
not voiced by the consumer panel.

Addition of 1.8% (w/v) fructose to the sample wines
resulted in significant increase in preference by both
panels of the sweetened white and red wines over their
respective controls (p<0.05). However, the panels did not
distinguish between the sweetened wine and the sweetened
wine to which emission volatiles had been added.

The results therefore suggest significant
contribution to the aroma/bouquet properties of both white
and red wines by the addition of fermentation emission
volatiles. Also, that sweetness seems a more important
attribute than added volatiles.

Selective Addition of Volatiles

In view of the results obtained above it was decided to
first examine the rough constituency of the volatile
stream emanating from the fermentation tank vents at a
subsequent crush using the same varieties of grapes.
Again, the wines were fermented dry. This study was done
in order to ascertain the results obtained when adding to
the wine only the fraction or fractions containing a
minimum of fusel oils since fusel oils, particularly
active- and isoamyl alcohols are generally considered to
impart a negative connotation to both aroma and bouquet of
brandies and wines (12).

Figures 6 and 7 show the pattern of emergence of
fusel oil components with respect to time for both the
white and the red wine fermentation. Based on such
emergence, volatiles were collected into two main
fractions: Fraction 1, prior to emergence of fusel oil
constituents; and fraction 2 comprising fusel oils and
constituents emanating toward the end of fermentation.

Results

Addition of volatiles of fraction 1 in similar ratios as
those used for the previous study to white wine (2X),

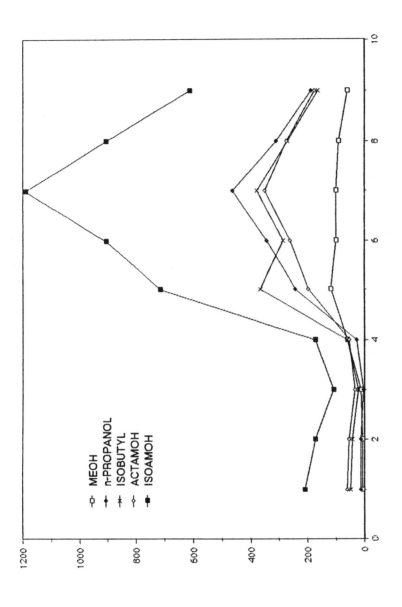

DAYS

PPM

Figure 6. White Wine Fusel Alcohols.

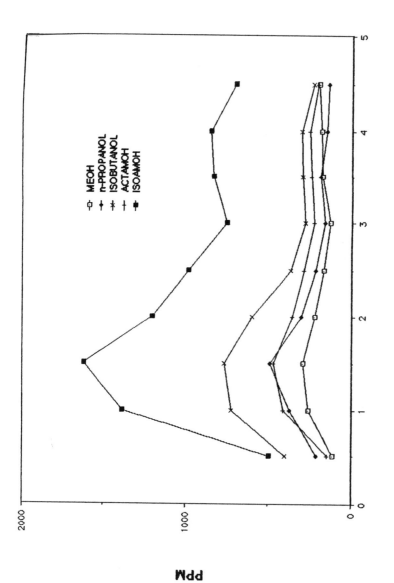

DAYS

Figure 7. Red Wine Fusel Alcohols.

resulted in a significant improvement, at $p<0.005$, in both intensity and acceptance of the resulting wine by both a trained panel (n = 20) and a consumer panel (n = 82). Similar results were obtained when fraction 1 of the volatiles from a red wine were added back at 4X ratio.

Addition of volatiles from fraction 2, which contained primarily fusel oil constituents, showed a significant increase in the complexity of the aroma/bouquet of both white and red wines. However, these wines showed a general trend toward less acceptance by both panels. The data for this trend however, does not show significance at $p<0.005$ level.

Conclusions

The results of our experiments over two consecutive seasons on the capture and addition of fermentation emission volatiles indicate that it is indeed possible to enhance the quality of wines by reincorporating the volatile fraction which otherwise would have been lost into the atmosphere.

This improvement is particularly significant when one considers the possibility of fermenting a white wine at the fermentation temperature commonly used for fermenting reds (ca. 27 deg C) and by capturing and reincorporating such volatiles in the manner described above. We have explored this concept for two crushes. Results for the first year of this study proved statistically inconclusive although a general trend toward improvement of wine quality was observed. Result of this years crush (1992) are not available as of this writing.

Should the results show improvement in overall quality of the wines thus fermented, then the energy savings in terms of refrigeration costs for wineries wishing to opt for this procedure to ferment white wines might be considerable. Savings might also be realized in operational expenses due to shorter turn around times and improved utilization of fermentation tanks ("cooperage").

It is not clear at this time if the sizeable equipment investment costs necessary to capture the volatiles might be considerably offset by wine quality improvement to warrant installation of such equipment. In addition, labor, maintenance and equipment cost required to fractionate the volatiles might be too onerous for implementation.

Acknowledgments

The authors wish to express their appreciation to the American Vineyard Foundation (**AVF**) and to the California Agricultural Technology Institute (**CATI**) of CSU, Fresno for their financial support. Our thanks also to Arthur Caputi, Jr. and to Mike Hardy for their review of this manuscript and for their helpful comments. We thank Christina Nasrawi for sensory evaluation and statistical analysis. Our thanks also to J. Scott Burr for assistance in winemaking and to Gwynn Sawyer Ostrom for assistance in preparation of this paper.

Literature Cited

(1) Rapp, A. In *Wine Analysis*; Linskens, H.F., Jackson, J.F., Eds.; Modern Methods of Plant Analysis; Springer-Verlag: Berlin, 1988, Vol 6; pp 29-66.

(2) Webb, A.D.; Muller, C.J. *Adv. Appl. Microbiol.* **1972**, *15*, 75-146.

(3) Rapp, A.; Mandery, H.; Niebergall, H. *Vitis.* **1986**, *25*, 79-84.

(4) Rapp, A.; Mandery, H.; Ullemeyer, H. Neuere Ergebnisse uber die Aromastoffe des Weines. Oenolog. Symp., Rome, 1984; pp 157-96.

(5) Rapp, A.; Mandery, H.; Ullemeyer, H. *Vitis.* **1984**, *23*, 84-92.

(6) Schreier, P.; Drawert, F.; Junker, A. Z. *Lebensm. Unters Forsch.* **1975**, *157*, 34-37.

(7) Hock, R.; Benda, J.; Schreier, P. Z. *Lebensm. Unters Forsch.* **1984**, *179*, 450-452.

(8) Breton, A.; Surdin-Kerjan, Y. J. *Bacteriol.* **1977**, *132*, 224-32.

(9) Eschenbruch, R. *Am J. Enol. Vitic.* **1974**, *25*, 157-61.

(10) Schmitt, A.; Curshmann, K.; Koehler, H. *Rebe Wine* **1979**, *32(9)*, 364-67.

(11) Laurent, M.; Lavin, E.H.; Henick-Kling, T.; Acree, T.E. Abstracted in *Am. J. Enol. Vitic.* **1992**, *43(4)*, 391.

(12) Reed, G.; Nagodawithana, T.W. In *Yeast Technology*; Van Nostrand Reinhold: New York, NY, 1991.

(13) Singleton, V.L. In *Wine Analysis*; Linsken, H.F.; Jackson, J.F., Eds.; Modern Methods of Plant Analyses; Springer-Verlag: Berlin, 1988, Vol 6; pp 173-218.

(14) Todd, D.F. Presentation on San Joaquin Valley Growth and Air Quality Impacts; Fresno, CA, April 7, 1988.

(15) Todd, D.F.; Catronovo, C.L.; Ouchida, P.K. Ethanol Emissions and Control for Winer Fermentation Tanks, Report #ARB/ML-88-027; California Air Resources Board Monitoring and Laboratory Division, Sacramento, CA, April, 1988.

(16) Muller, C.J.; Gump, B.H.; Fugelsang, K.C. Ethanol Emission II Project, Final Report; CSUF, VERC; Submitted to American Vineyard Foundation, California Wine Commission and Wine Institute, Nov. 15, 1988.

(17) Todd, D.F.; Castronovo, C.; Fugelsang, K.C.; Gump, B.H.; Muller, C.J. Ethanol Emissions Control from Wine Fermentation Tanks Using Charcoal Adsorption. A Pilot Study; Calif. Agric. Technol. Instit. #900705.

(18) Nasrawi, C.W; Wahlstrom, V.L.; Fugelsang, K.C.; Muller, C.J. Capture and Utilization of Emission Control Volatiles; Presented at Grape Day, Calif. State Univ. Fresno, Nov. 3, 1990. CATI Publication #901102.

RECEIVED March 30, 1993

ADVANCES IN HOME BREWING AND WINE MAKING

Chapter 13

Home Beer Making
Chemistry in the Kitchen

R. P. Bates

Food Science and Human Nutrition Department, University of Florida, Gainesville, FL 32611–0370

Home brewing, catalyzed by the Campaign for Real Ale in England; legalization in the U.S.; and a support network of brew clubs, associations, technical literature and suppliers, is undergoing remarkable growth. The subsequent steps - establishment of brewpubs and even microbreweries by ex home brewers - further popularize the practice. As with gourmet cooking, a strong chemistry/food science background is useful but doesn't guarantee brewing awards; art still plays an important role. However, an understanding of and appreciation for brewing chemistry and technology is usually acquired. What home brewers lack in technical expertise, they make up for with experience, innovation, enthusiasm and dedication; chemophobes are in the minority. The trends and practices described contribute importantly to the quality and diversity of beer now available in the U.S.

"A man who does not care about the beer he drinks may as well not care about the bread he eats.... Some people take their pleasures quickly, and swear loyalty to the same beer every day, but they miss much. The search for the perfect pint should last a lifetime. In the meantime, there is a classic style of beer for every mood and moment...." (*13*)

This quote quite elegantly explains the rationale behind home brewing. In the most general sense this avocation is the noncommercial manufacture of beer in a home environment (kitchen, garage, basement, etc) for personal consumption. Disciples range from highly skilled professionals to rank amateurs with no technical background in the many skills and sciences which contribute to brewing technology. The

0097–6156/93/0536–0234$06.00/0

common denominator is an interest in and commitment to the production and consumption of good beer.

An apt analogy is gourmet cooking where art and interest play a greater role than scientific and technical background. In both home cooking and brewing the results range from disasters to world class, with dedication, practice and learning from mistakes being most important.

Beer chemistry and technology is sufficiently intriguing and illustrative of the brewing process to merit special consideration. Chemistry has always been the central science and home brewing is a good example. There are constant reminders of chemistry's indispensable role thruout beer manufacturing and consumption. Chemists generally make good home brewers and contribute much to the field. Their involvement at the professional level is evident in other Chapters and ACS literature. It is, therefore, of interest to follow the homebrewing process from the perspective of the food chemist as distinguished from that of a commercial brewer.

Background

Home brewing has an interesting cyclic history (Figure 1). As the art evolved in agricultural communities (*17*) it went from the home to organized breweries, in the hands of craftsmen and eventually guilds. When people migrated, as to the New World, brewing again started in the home as a important feature of household self-sufficiency (*6*). The abundance of raw material, pioneering spirit and ethnic diversity resulted in the eventual establishment of home brewing-catalyzed small breweries thruout North America. Prohibition ended this positive development. Although home brewing survived and prospered during prohibition, beer quality was not the primary emphasis. Practitioners and advocates kept an understandably low profile; the sharing of information and popularization of the practice was severely dampened by the legal implications.

After Repeal and during the Depression, surviving breweries and new establishments commenced operations. However, by a curious quirk in Federal alcoholic beverage laws, home beer making was still prohibited (*22*). Raw material shortages during the 2nd World War and the economics of mass production and advertising subsequently caused a devastating shake-out in the brewing industry. Consequently, by the mid 1980's roughly 90% of all beer manufactured in the U.S. was produced by 6 major companies in Megabrewers with capacities exceeding 1 million barrels annually (31 million gallons). The unhappy result was a severe restriction in beer styles and versatility. Unless one patronized the imports or a few small surviving U.S. breweries, there was a grim monotony in choice. A similar situation was developing in England with industry consolidation threatening the strongly entrenched pub tradition.

Then commenced the beer revolution. In the early 1970's long suffering British beer consumers initiated "The Campaign for Real Ale" (CAMRA) which effectively reemphasized cask conditioned or "fresh" beer in contrast to the pasteurized, kegged or bottled product, as well as diversity in styles (*13*). The successful CAMRA emphasizing local and regional beer didn't reverse the tide of consolidation, but it did open an important niche for small breweries and brewpubs.

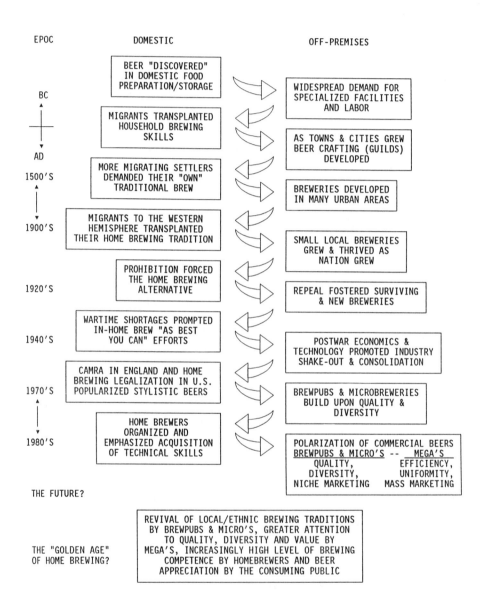

Figure 1. The Home Brewing Cycle

In the U.S. similar threats to quality beer were occurring and CAMRA was a highly relevant, gratifying model. With the legalization of home beer making by the federal government in 1979 there has been a proliferation in home brewing as reflected in brew clubs and associations, magazines, books and home wine & beer supply shops (22). A head of a household (over 21 years old) may prepare for personal consumption up to 100 gallons of beer annually (200 for a couple), provided the beer is not sold nor distributed indiscriminately. This represents almost three 12 oz bottles/day (slightly over the "healthy" limit of individual alcohol consumption). The 50 states and many counties still have a crazy patchwork of restrictive laws relating to home brewing and brewpubs which the respective associations are attempting to overcome.

Legalization paved the way for the survival of a number of traditional beer styles as well as the development of new ones. The evolutionary process has been equally beneficial as enthusiastic, competent, highly motivated home brewers started brewpubs - establishments which brew and serve their own beers on premises, some with limited outside distribution. Batches consist of about 10 barrels (1 barrel = 31 US gallons, 117.3 liters) each. An annual total of a few hundred to several thousands gallons of either their "Flagship" brands or special offerings, usually of remarkable quality in interesting, well executed styles are produced. The clever names and hilarious, often bawdy labels given to these beers illustrate the free-spirited imagination and enthusiasm of these brewers. There are now close to 300 Brewpubs in the U.S. with no signs of a slowdown in growth of numbers, beer volume nor quality. The next step, a microbrewery devoted primarily to off-premises sales with a volume of up to 60,000 barrels (1.9 million gallons) annually, have been successfully taken by a number of enterprising former home brewers. Thus, for the U.S. beer aficionado this is approaching the "Golden Age" of beer with home brewing to a large extent responsible for the vast improvement in beer diversity and quality now available in the U.S.

Table I indicates brewery categories and suggests a nomenclature. The scale ranges from homebrewers, decabreweries producing less than an individual's legal limit annually ~3.2 barrels to Gigabreweries - Companies with several plants each manufacturing more than 1 million barrels. The distinction between categories is blurred (8). Some original, so-called microbreweries now exceed 100,000, even approach 1 million barrels, and have national distribution. Whereas, a few megabreweries have primarily regional distribution. It is remarkable that with less than 10% of the market, brewpubs and micros account for over 90% of beer styles commercially available while homebrewers continue to fill every conceivable beer category niche and invent even more (3).

Although total volume of home, brewpub and regional microbrewery output matches, at most, the capacity of a megabrewery, at least the selection is now increasing and reversing the unfortunate post-prohibition trend. While more than 90% of beer drinkers seem irreversibly committed to the uniform, low flavored megabrewery offerings (thanks to superb technology and persistent, innovative marketing), those few percent who recognize and appreciate the true diversity inherent in beer styles need not depend upon foreign imports. Nevertheless, as in the U.S., international offerings also continue to improve in variety and quality, providing even more selection for beer connoisseurs.

TABLE I. BREWERY MAGNITUDES

CATEGORY DESIGNATION	DISTINGUISHING FEATURES	ANNUAL PRODUCTION IN BARRELS [a]
Home Brewery	Kitchen scale	3 - 7
Decabrewery	Advanced, non-commercial miniature brewery	3 - 7
Hectobrewery	Single brewpub, primarily on-premise sales	50 - 1,500
Kilobrewery	Brewpub Chain, on & off-premise sales	1,000 - 20,000
Decakilobrewery	Microbrewery, primarily off-premise sales	10,000 - <1,000,000
Megabrewery	Single brewery, regional or national distribution	~1,000,000
Gigabrewery	Company with multiple megabreweries, national distribution	>>1,000,000

[a] 1 Barrel = 31 Gallons = 117.3 Liters

An appealing aspect of home brewing is the fact that a practitioner doesn't have to understand the science to be successful. There is an element of art involved in following or developing a recipe; sensing and controlling those steps which influence beer quality and character; and finally evaluating, balancing, and fine tuning the most critical parameters to consistently produce fine beers for personal enjoyment or peer recognition. Overcoming uncertainty - outwitting nature - has appeal to both the scientist and nonscientist.

In the brewing process it is difficult not to become aware of the importance and interaction of agriculture, chemistry, physical chemistry, microbiology, engineering and other disciplines and their influence upon beer quality (5, 9). Individuals with little or no aptitude in science find the applied art of brewing pragmatic, useful, informative and entertaining. After dealing with water and starch chemistry; the physical chemistry of foams; the influence on flavor of certain natural constituents at parts per million (or less); hop isomerization and the dangers inherent in poor sanitation and yeast misuse, the non-scientist gains a healthy appreciation for scientific methods and values. Thus, home brewing popularization has value to the Chemistry and Food Science communities.

Nature is not benign and hazards to sound beer lurk in all facets of the brewing process. These can be overcome only by understanding and applying chemistry, microbiology, food technology etc; otherwise, disaster. This is an important message that all who personally brew beer - or prepare food, for that matter - should grasp. Thus, with few exceptions, home brewers are one step beyond the general public and particularly food activists, who personally have or promote an irrational fear of chemicals in food. Although chemophobes exist and are vocal, they are a minority in the home brewing fraternity. If they make good beer in spite of chemophobia, fine; they are tolerated to a greater extent than they reciprocate. In brewing one deals with a substance (alcohol) which, when misused, is a tangible health risk, in contrast to trace or non existent substances in the normal food supply. Gratifying, moderate alcohol consumption seems to have a definite, yet ill-defined protective effect against a number of human illnesses (*14, 27*).

Home Brewing Practices

The humorous, exaggerated stereotypical image which accompanies home beer making is still persistent. Clandestine (moonshine-type) operations and exploding bottles, while part of the heritage, are now exceptions to the rule. The comprehensive, growing literature on home brewing and brewpub brewing adequately addresses the intellectual and practical needs of both beginner and expert (*22*). There are few major cities which don't have at least one active home brewer club, frequently encouraged by local brewpubs or home wine & beer supply stores. There are over 200 homebrew clubs in the U.S. and their numbers are increasing even faster than those of brewpubs. Many clubs are members of larger organizations which hold regional and national meetings with well attended technical sessions, industrial exhibits, tastings and competitions.

The American Homebrewers Association provides an extremely important service in their Beer Judge Certification Program (*4*). Uniform beer evaluation criteria are established, training is provided and judging competitions are standardized

(23). The quality and diversity of styles at these competitions do justice to the science of brewing; they are as structured, thorough and organized as any professional wine competition. These activities have generated a number of texts and journals on the subjects. Their technical quality is gratifyingly high and the ideas presented are often quite innovative, reflecting a good understanding of chemistry, microbiology, brewing mechanics and agriculture. Although early homebrewing literature contained a few inflammatory outbursts about chemicals in commercial beer (is there anything else?) a more sensible, informed tone now exists. Some publications match the depth and coverage of scientific literature. Courses, short courses and workshops on home brewing are also presented by some universities as a part of degree or community education programs.

Beer supply houses provide the necessary raw material, equipment, directions (literature) and encouragement. Major items are kits, consisting of canned hopped or unhopped malt extract; accompanying dried or liquid yeast packets; various specialty dried malts, adjuncts and all forms of hops. Canned malt comes from a number of foreign, many British, and domestic malt houses. The selection of kits and accessory materials/equipment is growing. So the gourmet brewer has direct (or mail order) access to even the most exotic materials. Suppliers and brewers now recognize the need for quality; proper handling, storage and inventory control are the rule. Previously, items requiring refrigeration or cool storage such as hops and yeast were often abused.

Thus a strong moral, technical and logistic support network exists for the home brewer at all levels of expertise and commitment. Perhaps not unexpectedly, commercial breweries - Brewpubs, Micro-, Mega-, even Giga-breweries support this network; recognizing that diverse, quality beers are in the industry's best interest.

Table II indicates a few of the reasons for the popularity of home breweries and some of the obstacles. The chemist has an additional challenge - combine art and science to excel. Interestingly, this has not yet transpired. Chemists have their share of home brewing awards and recognitions, but they do not dominate the field. Art still plays a major role.

A highly simplified beer making schematic which broadly covers the entire process is shown in Figure 2. Table III indicates equipment needs and critical control points for home brewers. The beginning home brewer can eliminate a number of the time consuming, complex steps by simply employing a hopped malt extract with a packet of dry beer yeast, adding sugar and following directions on the can as outlined in the diagram. As experience and confidence are gained the home brewer can then move more into the mainstream and eventually to all grain brewing with its limitless combination of ingredients, mashing/hopping schemes and fermentations regimes (28, 12). In fact, it is possible to establish a complete home "mini-micro-brewery" capable of performing practically all the operations performed in a commercial brewery. Home and Brewpub brewers with ingenuity and the assistance of equipment suppliers have devised novel ways of accomplishing practically all essential brewing unit operations, often by the clever use of common, inexpensive household and hardware equipment. Such schemes attest to the resourcefulness and imagination of home brewers.

Homebrewing is not amenable to strict cost accounting (18). If it were, some practitioners would be better off partaking of the premium offerings from an upscale

TABLE II. THE PRO'S & CON'S OF HOME BREWING

ADVANTAGES

- Simple, well-tested recipes exist

- Wide variety of raw material and equipment readily available, supplemented by common kitchenware

- Can be done on a small scale (~ 1 to 5 gallons), at minimal expense

- Extremely high quality or exotic beers (unavailable, nonexistent or relatively costly) can be produced

- As with gourmet cooking, ample opportunities for creativity, individualism

- Established network of homebrewers promotes experimentation, competition, brewing skills and comradarie

- Practitioners gain useful insights into the role of chemistry, food science and technology in foods (counteracts chemophobia)

- A socially rewarding hobby

DISADVANTAGES

- Time consuming, ~1 month lag time

- Ties up space, some operations tedious & messy

- Critical steps - temperature control and analyses can get complicated

- Dangerous - if bottled beer overcarbonated

- Uncertain legal status in several states

- Raw materials can be somewhat costly and variable in quality

- Increasingly, quality beers in all styles commercially available

- Practice may be addictive - reduces enjoyment and appreciation of mass-marketed beers

---

(.. ..) Denotes Beginning Homebrewer's Options

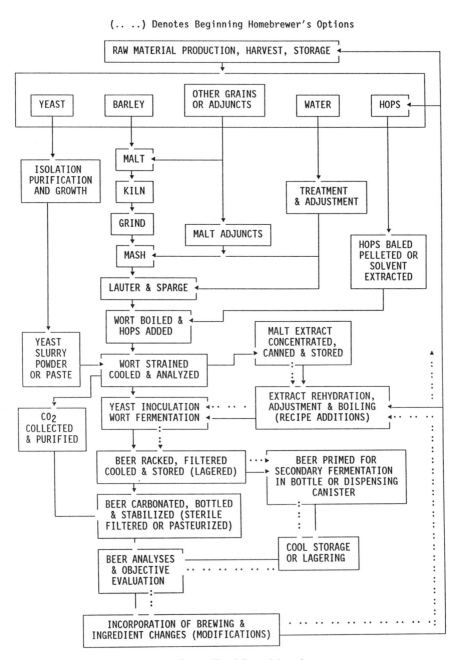

Figure 2. Generalized Beer Manufacture.

brewpub or microbrewery. Practices can be as cheap as < 25¢/12 oz bottle or as expensive as home brewer tastes demand, particularly high, if personal time and capital costs are factored in. However, as a means of developing useful skills and as a source of personal pride and enjoyment, homebrewing is a rewarding hobby.

The Brewing Process

As suggested by Figure 2, there are many paths thru this scheme, depending upon the interest, skill, commitment and resources of the home brewer (*22*). All involve a fascinating balance of theory and practice with examples touching on most aspects of food science and technology and chemistry.

Raw material - The German Beer Purity Law of 1516 "Reinheitsgebot" defined the sole ingredients for beer (Table III) (*16*). While this standard still serves as an ideal model with modifications (i.e. recognition of yeast, use of other malts and unmalted barley), the home brewer now has bewildering array of options and alternative materials by which to pursue the avocation.

Barley is the traditional but not exclusive source of malt. The discovery of malting, by which steeped (rehydrated), viable grains start to sprout, thus producing the enzymes essential for starch conversion to fermentable sugars, was an important achievement of civilization and intrinsically linked to the development of agriculture. This was perhaps the first biotechnology and is still under active investigation (*20*).

Comparatively few brewers, commercial or amateur, produce their own malt. Malt houses provide a range of malts suitable for all styles and qualities of beer. The malts vary in their enzyme profiles, colors and flavor contributions. These attributes are influenced by the barley cultivar, malting and kilning steps. Kilning is the drying and flavor/color development process which stabilizes the malt. The kilning regime can either be gentle (~50-60°C) to maintain a high, selective enzyme activity - primarily amylolytic with some proteolysis (to enhance foam stability) - or proceed at a high temperature (>150°C) to inactivate all enzymes but produce a host of Maillard reaction products and accompanying desirable dark color and distinctive toasted flavors. A mixture of malts contributing both hydrolytic enzyme activity and flavor/color is generally employed.

Unlike vintners who frequently either grow their own grapes or closely supervise grape cultivation, brewers are usually not involved in barley cultivation nor malting. Successful brewers do provide malt houses with clear specifications regarding malts and pay close attention to malt quality. As in enology, final product quality is defined by the starting materials.

Water chemistry is of considerable importance in brewing. Beer styles depend upon the right water - fairly soft for Pilsen, hard for English Ale and various salt balances in between. Calcium and pH especially influence enzyme activity and polyphenol solubility (*5*). For the beginner, tap water without obvious off-flavor, high mineralization or chlorination is adequate. Brewing recipes often provide suggestions for modifying water to match beer style (*25*). Thus, the home brewer should have some idea of starting water quality and hardness.

Hops may be thought of as the spices and condiments of beer. Along with some preservative effects (less critical in the presence of good sanitation and refrigeration), the bitter flavor and spicy aroma of beer come from hops. Natural

TABLE III. HOME BREWING OPERATIONS, MINIMUM EQUIPMENT NEEDS AND CRITICAL CONTROL POINTS

OPERATION	EQUIPMENT	CRITICAL CONTROL POINTS
Malting	Kitchenware	Seed viability & moisture, humidity, temperature
Kilning	Oven, thermometer	Temperature, seed distribution, time
Grinding	Coffee or grain grinder	Medium coarse grind
Water addition	Volumetric measure	Quantity, quality, hardness
Mashing	Pot	Temperature - time profile
Lautering & sparging	Strainer	Husk size & distribution, filter bed porosity
Wort boiling & hops addition	Kettle, scales, hydrometer	Hop quality, quantity, time of addition
Wort cooling & settling	Skimmer, strainer	Sanitation (throughout) ↕
Yeast addition	Plastic/glass container	Yeast type, viability, purity, quantity, wort temperature
Primary fermentation	Water seal, cool location	Even temperature (~7-24°C), Anaerobic conditions
Racking	Siphon hose & alternate container	Complete fermentation, settled sediment, aeration
Priming	Volumetric measure	Quantity of residual & added sugar, priming yeast viability
Bottling	Bottles, caps & capper	Bottle & seal integrity, sanitation
Aging	Cool, dark location	Temperature (~4-24°C), time
Evaluation	Discriminating palate, knowledge of beer styles	Inappropriate serving temperature, unfamiliarity with beer quality, overconsumption

product chemists have made and continue to make important contributions to brewing. There is even more complexity in hops than in malts (*5, 16*). The transformations which occur before, during and after the brewing process are appreciated but not fully understood. Hop cultivars and chemistry are important research topics, as indicated in ACS literature.

While hops play a supporting role in standard commercial beers, they are more appreciated in many styles of premium and home beers (*2*). Home brewers with little chemistry background follow and understand in a practical sense terms like "isomerization", "α and ß acids", "humulene" and "bittering units" which are part of brewer's language. Furthermore, hop extractions by organic solvents, liquid and supercritical CO_2, which improve and simplify hop utilization procedures, has meaning to non-engineers.

Yeast, the "bag of enzymes" which carry out the deceptively simple Gay-Lussac equation (Figure 3), are well appreciated by home brewers. Actually, the myriad of anabolic and catabolic steps accompanying yeast metabolism and affecting beer character is the keystone of life processes in all organisms, including humans (*8, 11*). The development of special beer/ale dry yeast of high stability and purity eliminates the tedious operation of yeast cultivation from agar slants. For the more adventurous, liquid yeast, mixed cultures, yeast reuse and even isolation of strains from some unpasteurized commercial beers is possible (*1*). Few home brewers depend upon wild yeast or fail to employ strict sanitation precautions thruout the brewing process. Otherwise, spoilage is a dramatic reminder of the microbiological aspects of brewing. If the average home cook were as careful as most home brewers regarding sanitation and housekeeping, the home-induced incidence of food-born disease would be much lower.

Increasingly, home brewers are turning to all grain brewing or a combination of malt extracts and dry malt to increase beer complexity and quality. This requires attention to malt combinations, enzyme activity and the mashing regime. While requiring more time, effort and equipment, dry malt is less expensive than concentrated extracts and mashing provides greater versatility. However, the wide choice of extracts by themselves or in combination with dry malts can yield excellent beers.

The mashing procedure dictates to a large extent beer character. The mashing temperature is programmed either by adjusting the heating regime in several steps (infusion) or by removing a portion of the mash for boiling and remixing (decoction). This sequence of mash temperatures and times selectively favors the action of the various malt enzymes including α and ß amylase and proteases. During mashing malt adjuncts (sources of starch or special character) can be added to take advantage of malt enzymes. In general, well modified malts with high diastatic (starch digesting) activity yield worts high in fermentable sugars - alcohol potential. Whereas, less modified malts or mixes including enzyme-inactive malts or shorter mashing regimes leave considerable nonfermentable carbohydrates (partially hydrolyzed starch fragments) which contribute to malt flavor and beer body.

Malt adjunct use is controversial. These inexpensive sources of starch - corn, rice, even glucose syrup - can replace the more costly barley malt at up to 60%, or more, if industrial amylolytic enzymes are also employed during mashing. Such adjuncts contribute little to flavor or body but are effective diluents in lighter style beers. Unfortunately, competitive pressures and mass marketing dictate that adjunct use is more common and probably increasing, to the detriment of beer quality. Germany recognized the relationship almost 500 years ago with Reinheitsgebot (Figure 4, presently a point of contention in European Community trade negotiations), as do most homebrewers, brewpubs and micros.

The subsequent lautering and sparging operations by which the sweet wort is strained thru and leached from the spent grain bed can be performed with pots, pans and strainers. Home brewing supply houses now stock simple false bottom plastic pails with a heating element and appropriate plumbing to accomplish these potentially messy steps.

Independent of the complexity of the brewing scheme, the sweet wort (solubles extracted from malt and hops) must be adjusted for fermentation and boiled. For debatable reasons, many homebrew recipes call for corn sugar (glucose) in lieu of sucrose (22). Wort boiling accomplishes several important operations. It pasteurizes the wort, develops the bitterness of added hops and precipitates or coagulates protein and colloids. As with malts and mashing, there are many hopping schemes involving various hop cultivars, quantities and time of addition. Hop resins added early in the 60 to 90 minute boil will be isomerized to produce a high bitterness level. Hops added later, particularly at the end of the boil, provide less bitterness but the volatiles are not steam distilled; hence they contribute hop aroma. Whole hops or pellets can be supplemented with or replaced by extracts, oils and even aqueous essence to provide home brewers with practically the same hopping alternative as their commercial counterparts. The use of hops is still an inexact science and a fascinating exercise in creative brewing.

The boil-induced particulates (Trub), removed by centrifugation commercially, can be skimmed off or settled out as the wort cools in the home. At this point sanitation becomes especially critical. Sterility isn't possible nor necessary, but the wort must be handled carefully to avoid undue contamination and cooled to below yeast-damaging temperatures ($<45°C$). Fortunately, a vigorous inoculation of the appropriate beer yeast will usually dominate the fermentation. With the advent of reliable sanitation, temperature control and yeast purification, the distinctions between top fermenting (ale) and bottom fermenting (beer) yeast is less critical. Yeast strain influences beer style to some degree, but less than other recipe factors - provided the yeast is viable and added in sufficient quantity ($\sim10^7$/ml of wort). Hermetically packaged dry beer yeast stores well under refrigeration, but deteriorates fairly rapidly at or above ambient temperature. The same holds for hops, so both should be refrigerated and employed fresh in homebrewing.

The alcoholic fermentation is an anaerobic, exothermic process, requiring cooling (Figure 3). Fortunately, the high surface area/volume ratio of 5 gallon carboys used in homebrewing easily permits dissipation of this heat. However, fermentation temperature should be between about 5 and 25°C, depending upon style. At the extremes the fermentation proceeds slowly or sticks (ceases) or is too rapid and beer quality suffers. Conscientious homebrewers deal with temperature control by cooling with turned up air conditioners, old refrigerators, and even insulated walk-in chambers.

In the interest of sanitation and alcohol production, it is critical to insure anaerobic conditions during the fermentation. At yeast inoculation, dissolved oxygen

$$C_6H_{12}O_6 \xrightarrow[\text{+ Nutrients}]{\text{Yeast}} 2\ CO_2 \uparrow + 2\ C_2H_5OH + 209kJ$$

| Fermentable Sugars | $- O_2$ (Anaerobic Environment) | Carbon Dioxide | Ethanol | Heat |

(Primarily mono & disaccharides)

Figure 3.   The Gay-Lussac Equation

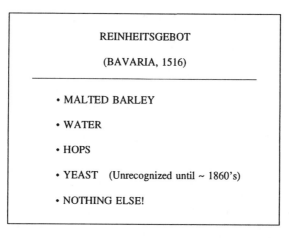

REINHEITSGEBOT

(BAVARIA, 1516)

- MALTED BARLEY

- WATER

- HOPS

- YEAST   (Unrecognized until ~ 1860's)

- NOTHING ELSE!

Figure 4.   The German Beer Purity Law - Permitted Ingredients

in wort allows limited aerobic growth of yeast cells and the build up of sterols, essential for subsequent metabolic processes (19) during anaerobic fermentation. Later, vigorous CO_2 evolution effectively purges residual air and provides a barrier to ambient oxygen diffusion into the wort. The fermentation then should be protected from air and contaminants by a simple U-shaped water-filled fermentation lock. From this point on oxygen is an undesirable component in beer finishing and aging reactions and air contact is to be minimized.

At the completion of fermentation commercial breweries have the advantage of equipment and facilities for centrifugation, filtration, controlled temperature aging (lagering), pressure carbonation, and pasteurization or sterile filtration. In contrast, all but the most elegantly equipped home brewery rely upon racking (decanting or siphoning beer from lees, i.e. sediment), holding the beer at available storage temperature and in-bottle carbonation. In-tank, soft drink canister pressure carbonation (using, hopefully, only food grade CO_2) is becoming more widespread (24). However, home beer is usually primed with added sugar prior to bottling. Residual yeast convert the sugar to alcohol and CO_2 to develop required carbonation.

This is clearly the most dangerous step in homebrewing. If the beer contains much residual fermentable sugar, too much sugar is used in priming, or defective bottles are filled, an explosion hazard exists. Since 0.1% fermentable sugar can generate about 5 psi at 25°C in a closed container, small errors in misjudging the completion of fermentation or priming amount can result in extreme bottle pressures. When combined with high storage temperature or agitation during transportation, an innocuous looking home beer is a "glass bomb". Some home brewers use plastic soft drink bottles which are safer, functional for short term storage, but less aesthetic.

Even when done correctly, in-bottle priming leaves a detectable yeast sediment, which can be minimized by carefully decanting upon serving. The sediment can be avoided by a dialysis system which separates yeast from beer or in-bottle pressure carbonation devices described in home brewing literature (7). Of course, neither yeast sediment nor the protein-tannin hazes which can develop influence beer flavor. In fact, such defects are a positive quality attribute in several superpremium commercial beers.

Aging (lagering) at cool temperatures (4-15°C) for a few weeks to months allows home beer to develop carbonation and mellow; some of the harsher flavor notes subside and the yeast sediments. The duration of maturation and the subsequent shelf life depends upon storage temperature, beer style and subtle factors. In the absence of pasteurization, homebrew may not store as long as commercial beers. But, with attention to quality in formulation, proper brewing procedures and reasonably cool storage, it need not and will not last as long.

Evaluation

While lack of a technical background or experience need not be a deterrence, there is one very important characteristic which the novice home brewer should possess; that is - enjoy beer. A general familiarity with beer types, styles and brands as well as an appreciation of beer quality attributes and how to determine them are useful. Popping a "cold one" on a hot day or consuming prodigious quantities of a prominently advertised national brand is not a proper introduction to fine beer.

Consumption of beer near its freezing point does a grave injustice to quality beers which should be consumed at 45 to 60°C, depending upon style. The converse applies to bland or mediocre beers. Quantity counts for little and actually detracts from real beer appreciation. Attention to quality and development of a discriminating palate by continual tasting (not guzzling) is essential. Fortunately, it is fairly easy to sample a wide selection of beer styles by visiting specialty shops, brewpubs, microbreweries, home brew clubs or homebrewing colleagues. Talking with brewers and sampling their wares is a worthwhile, pleasant training and an effective way to learn beers and brewing technology. As suggested by Jackson (*13*), this is an enjoyable, endless pursuit. Brewers - home & otherwise - are highly individualistic about their preferences and interpretation of styles. This comes from experience, not beer commercials. For those so inclined, there is a popular, comprehensive beer judge training and certification program offered nationally (*3*).

An important part of the homebrewing process is evaluation. As emphasized, just drinking beer ice cold is not thoughtful evaluation and analysis. Well defined quality criteria, obtained by experience and attention to detail is valuable (*13*). Obvious defects such as under or over carbonation, inadequate or excessive foam, particulate matter, off flavors, harsh acidic taste (due to lactic acid bacteria, rarely acetobacter) simply reflect gross errors in procedure and are easily recognized and corrected in the next batch. Home brew clubs and beer-knowledgeable colleagues provide a vital forum for homebrewers to evaluate and compare their efforts. Formal competitions with certified beer judges further reinforce quality expectations. Positive (or even negative) feedback obtained here and from brewpubs and microbrewery offerings, which set a high standard, combined with better raw material and equipment, have all contributed to improving the consistency, diversity and quality of home beer.

The two biggest shortcomings of home brewers seem to be:
(1)   Not trying hard enough - either being satisfied by inferior beers or not learning from mistakes.
(2)   Getting too fancy - embarking upon exotic formulations or incorporating too many variables at one time with little thought to or understanding of the confounding effect of these parameters.

"Far out" beer styles can be interesting and even palatable, but until a home brewer can consistently produce sound, recognizable beer, the efforts of experimentation are misplaced. Nevertheless, if a home brewer wishes to experiment wildly (no matter how extreme the concoction, provided it's not toxic), an individually designed, crafted and executed beer makes an elegant personal statement; sometimes weird, but definitely personal.

The Future and Challenges of Home Brewing

Barring the debacle of prohibition, home brewing popularization will undoubtedly increase. As government at all levels looks to "sin" taxes and, if the megabreweries continue in their cost-effective, lucrative pursuit of the hearts and minds (certainly not the tastes) of the mass beer market, the home brew alternative will become more attractive. The brewpubs and micros will continue to play a key role by setting high standards for beer quality and diversity.

Molecular biology has enormous potential to provide even better raw materials. Malt, adjuncts, hops and yeast all have their limitations which are amenable to genetic engineering techniques (26). Thanks to their support network, home brewers can rapidly take advantage of innovations in raw materials and techniques. Capital-intensive equipment such as centrifuges, ultrafilters, reverse osmosis units and temperature control systems are more difficult to acquire and some, i.e. continuous fermenters (15), are not necessary. However, enterprising home brewers and suppliers are likely to develop small scale, relatively inexpensive solutions to such operations.

There is one area demanding attention from home brewers and which may influence beer character somewhat. Emphasis is generally on full malt beers with body and a high flavor profile. To this tradition "lite" low calorie beers are an anathema and non-alcoholic beer even moreso. Yet there are beer aficionados who for physiological, philosophical or psychological reasons cannot partake of traditional beers. The giga-breweries are addressing this demand with some success with bland, uniform offerings. Although this approach may seem opposite to that of the home brewer, with proper attention to malting, mashing, hopping and fermentation it is possible to produce flavorful, unique low calorie beers. Non-alcoholic beer is more difficult (10, 21). Flavor development calls for a fermentation. The next step - alcohol removal without flavor stripping - requires special techniques and equipment, such as distillation with essence add-back, ultra-filtration or reverse osmosis. When and if the home brewing community recognize a need for such products, the hardware and procedures will follow.

Despite appreciable increases in the quantity and quality of home beer, there is another formidable challenge facing home brewers and their commercial kindred spirits. Fine beer is not held in high regard in the U.S. Connoisseurs of gourmet foods and fine wines will still generically order "a beer", although they put considerable thought into selections from a menu or wine list. This occurs despite the existence and availability in the U.S. of beers which match in quality and diversity well respected wines. Public attitudes would have to change if more than a few percent of beer consumers are to learn to appreciate the variety of styles inherent in beer.

As the fraternity (and increasingly sorority) of homebrewers grow and the level of their sophistication and popularity increases it is possible that their influence could cause the Megas and Gigas to consider diversifying the styles and quality of their offerings. Such decisions would be based upon economic considerations, since the level of understanding and application of brewing chemistry and technology is already exceptionally high. In fact, it is more difficult and more of a technical challenge to consistently and economically produce a uniform mass-market beer in large volumes than to produce small quantities of high quality beer in various styles. It simply isn't worth the effort to make and merchandise an upscale product, unless more beer consumers demand a variety of beer styles in high quality and at reasonable prices. However, the average American gets the beer he/she deserves as determined by market demand. It is up to the home brewing community and their support network to turn the U.S. from a beer consuming (guzzling?) nation to one where fine beer is held in high regard and expected, if not demanded by the majority. Such is still the fortunate case in a few North European countries (*13*), for which all beer connoisseurs should be grateful.

There is cause for optimism.   In the same sense that the science and technology of chemistry benefits enormously from its extremely interdisciplinary nature, the broad constituency of homebrewers represents practically all professions and perspectives.   The bringing together of such a diverse group certainly complicates but ultimately enriches the avocation.

Thus, the future looks bright for home brewers and their mentors - the brewpubs and micros.   In addition, parallel trends in Europe and eventually other regions will result in a revitalized brewing tradition involving old and new styles, limited only by the imagination of the practitioners.   Whether these positive developments will encompass more than a small fraction of average beer consumers or attract the attention of the megas and gigas is uncertain. Nevertheless, involvement of chemists is a welcome development.   When sampling outstanding, worldclass beers, if one (or better yet, one's colleagues) can truly say your product is in the same league, your brewing and applied chemistry skills have been put to good use.

Literature Cited

(1)   Am. Homebrew Assoc. Yeast and Beer: Special Issue, *Zymurgy* **1989**, 12(4).

(2)   Am. Homebrew Assoc. Hops and Beer: Special Issue, *Zymurgy* **1990**, 13(4):6-60.

(3)   Am. Homebrew Assoc. Traditional beer styles: Special Issue, *Zymurgy* **1991**, 14:(4).

(4)   Am. Homebrew Assoc. Beer Entusiast 3:(1). in *Zymurgy* **1992**, 15:(1).

(5)   Briggs, D. E., Hough, J. S., Stevens, R. and Young, T. W. Malting and *Brewing Science*; Volume I and II. Chapman and Hall: London, New York. **1982**.

(6)   Brown, S. C. Beers and wines of Old New England. *Amer. Sci.* **1978**, 66:460-467.

(7)   Daniels, S. How to build a simple counterpressure bottle filler. *Zymurgy* **1990**, (13):36-38.

(8)   Edgar, D. Small break for small brewers. *New Brewer*, **1991**, 8(1):30.

(9)   Eskin, M. Biochemistry of food processing: Brewing. Ch 6 in *Biochemistry of Foods*; Academic Press, San Diego, CA **1990**.

(10)   Fenton, M. S., Kavanagh, T. E. and Clarke, B. J. Recent developments in brewing technology. *Food Technol Aust.* **1988**, 40:132-134, 148.

(11)   Fix, G. *Principles of Brewing Science*; Brewers Publications: Boulder, CO. **1989**.

(12)   Hardwick, W. A. Beer. Chapt 3 in *Biotechnology*, Vol. V, Food and Feed Production with Microorganisms. Reed, G., Ed. Verlag Chemie. **1983**.

(13)   Jackson, M. *The New World Guide to Beer*; Running Press: Philadelphia, PA, **1988**.

(14)   Kannel, W. B. Alcohol and cardiovascular disease. *Proc. Nutr. Soc.* **1988**, 47:99-110.

(15)   Lommi, H., Gronqvist, A and Pajunen, E. Immobilized yeast reactor speeds beer production. *Food Technol.* **1990**, 44:128-133.

(16)   Macleod, A. M., Beer. Chapt 2 in Microbiol Technology, Pepper and Pearlman, eds, Academic Press, **1979**.

(17)   Maytag, F. Sense and nonsense about beer. *Chemtech* **1992**, 22:(3)133-141.

(18)   McCrady, R. Love or money? Why we brew. *Zymurgy* **1990**, 13(2):31-32.

(19)   Munoz, E. and Ingledew, W. M. An additional explanation for the promotion of more rapid, complete fermentation by yeast hulls. *Am. J. Enol. Vitic.* **1989**, 40:61-64.

(20)   Ogundiwin, J. O. and Hori, M. O. Development of stout from sorghum malt. *Food Sci. Technol.* **1991**, 24:182-185.

(21)   Owades, J. F. Preparation of alcohol-free barley malt-based beverage. U.S. Patent 4,765,993, August 23, 1988.

(22)   Papazian, C. *The New Complete Joy of Home Brewing*; Avon Books: NY, NY, **1991**.

(23)   Pfeiffer, W. Tasting techniques: Setting the stage for effectively evaluating beer. *Zymurgy* **1987**, 10(4):15-17.

(24)   Rager, J. Soda key draft systems: Better than bottles? *Zymurgy* **1989**, 12(5):24-27.

(25)   Rodin, J. and Colon-Bonet, G. Beer from water: Modify minerals to match beer styles. *Zymurgy* **1991**, 14(5):28-32.

(26)   Russell, I., Jones, R. and Stewart, G. The genetic modification of brewer's yeast and other industrial yeast strains. Ch 11 in *Biotechnology in Food Processing*; Harlander, S. and Labuza, T., Eds; Noyes Publ. **1986**.

(27)   Shaper, A. G., Wannamethee, G. and Walker, M. Alcohol and mortality in British men: Explaining the U-shaped curve. *Lancet* **1989**, (8633)336.

(28)   Westermann, D. H. and Hinge. Beer Brewing. in Prescott and Dunn's Industrial Microbiology, 4th Ed. AVI Publ Co.: Westport, CT, **1982**.

RECEIVED February 8, 1993

Chapter 14

Home Wine Making

Effect of Societies, Retail Outlets, and Competitions on Wine Quality

Gerald D. Cresci

Cresci Vineyards, 11746 Giusti Road, Herald, CA 95638–9726

Historically home winemaking has been a part of winemaking production in the United States. Those who participate in this process represent a sizable segment of the wine consuming public. Frequently individuals speculate as to whether the quality of homemade wines has improved over the last few years. Many factors contribute to the quality improvement of homemade wines. The most important ones include wine competitions, wine clubs, and winemaking supply shops. Homemade wine competitions are conducted on a national, state, regional, and local basis. Written evaluations from these competitions to the home winemakers provide the producers with immediate feedback regarding the quality of their wines. Clubs provide members with a communication network to learn about wine and how to improve them. Winemaking supply shops are available on a day to day basis as an immediate source for advice on how to solve winemaking problems. Formal educational institutions at this time are not a direct factor in the improvement of homemade wines. Given the limitations of home wine production a majority of home winemakers continue to produce acceptable wines.

Is home winemaking improving? In attempting to answer this question, I have drawn upon my own resources. There are no scientific facts to validate any conclusions presented. My resources are simply observations and experiences studying about winemaking, observing others make wine, and just being a part of the home winemaking history for the last sixteen years. Most of this winemaking history is shared with my wife Nellie, who has been a partner in all my winemaking activities. Our wine partnership began when we enrolled in an adult education class on viticulture. In the fall of 1974, Dan Pratt, a Sacramento nurseryman taught an adult education viticulture class in one of the local high schools. Pratt is a certified nurseryman with viticulture experience in Napa Valley. He currently works for a large Sacramento nursery and writes a weekly column on gardening for the "Sacramento Bee."

0097–6156/93/0536–0253$06.00/0

One of our first wine educational experiences was at the Napa Valley Wine Library Association's "Introduction to Wine Evaluation" Seminar in 1973.

Other classroom experiences include those at the University of California Davis Extension and The U.C. Cooperative Extension Service Classes. The latter classwork was two extensive viticulture courses offered through the San Joaquin County Farm Advisors Office.

Prior to making wine we attended a home winemaking class conducted at a local fermentation shop. The basic text for this class was <u>Enjoy Home Winemaking—A Guide for the Beginner</u>.

In 1975 we started to make wine and have continued to do so until this date. Our experience with making wine led eventually to the purchase of ten acres in the southern part of Sacramento County, and in the Lodi BATF Wine Growing Appellation. On this property we planted two acres each of French Colombard, Chenin Blanc, Cabernet Sauvignon,Petite Sirah, and one acre of Sauvignon Blanc. This farming activity and planning was undertaken to provide grapes for home winemakers.

Simultaneously with these events was our participation in home winemaking club activities. In 1976 we became aware of the Sacramento Home Winemakers, Inc.(SHWI), and joined this winemaking club. One thing lead to another and in 1977 my wife was asked to served as club secretary and was elected treasurer in 1978 and 1979. I became President in 1978 and served as Treasurer from 1980 through 1987.

Club involvement pushed me into wine competitions. Entering wine competitions led to working in them as a volunteer and ultimately to organizing wine competitions. The following list represents some of these home wine competitions and responsibilities:

* California State Fair: Judge's Clerk, one year; Assistant Co-Chair, two years; and Competition Chair, eight years.
* Sacramento Home Winemakers, Inc., May Jubilee Competition Chair for eleven years.
* Lodi Spring Wine Show, Home Wine Competition Chair for three years.
* Sacramento Wine Festival, Home Wine Competition Chair for two years.
* Home Wine and Brew Trade Association, International Homewine Competition Chair for one year.

Since 1987, I have been the California State Fair Commercial Wine Department Chair and have the administrative responsibility for the commercial wine judging and the home wine and home brew competitions. Part of the State Fair responsibility included the development of a wine judges' certification test. The test was administered to increase the "pool" of judges needed for the commercial wine competition. This experience was used to assist the Home Wine and Beer Trade Association in developing their homewine judges' certification test.

Many invitations have come to me to judge in homewine and commercial wine competitions. For the last five years I have had the good fortune of judging in approximately six home wine competitions each year.

The above information has been cited to establish my credentials as an observer of home winemakers and the quality of their wines. If you ultimately disagree with these observations I can only provide a disclaimer that the observations may be faulty or were misinterpreted.

Gift, Experience or Knowledge

How does one become a good winemaker ? Is the ability to make a good wine a gift, or is a good wine made by practice and the application of knowledge? Many times I've seen a winemaker defy all common sense and end up making a good wine. This doesn't happen often, but it does happen.

More frequently home winemakers will follow a procedure, or if you will, a recipe and never deviate from it. They will consistently make an acceptable wine. Some home winemakers are uncomfortable with anything but a strict adherence to a procedure. Invariably they will make an average wine and occasionally an exceptional one.

One advantage a home winemaker has is the opportunity to experiment. To experiment, one must have the knowledge of the tools and the ability to manipulate them to the best advantage. Good practices, applied properly, can maneuver a wine in the desired direction. All other factors being equal, the manipulation can result in different and distinct types of wines. The manipulation can lead one in the direction of developing a preferred personal wine type.

Experimentation in winemaking necessitates good record keeping. Success encourages winemakers to specialize. Some home winemakers make excellent red wines, but rarely have such success with white wines. This phenomenon also occurs in commercial wine production. There is one winery that consistently makes an outstanding medium-dry Chenin Blanc, but can rarely produce an acceptable dry one. Some home winemakers will carve a niche with a parsley or dandelion wine and never make anything else.

A few home winemakers make wine for their own consumption and care little about the quality. Some have so little experience in evaluating wine quality, they can't recognize a bad wine. A few years back, a local community college conducted a Wine Expo, which featured wine, food and crafts. The SHWI had a booth at the event. One visitor came to our booth and claimed to be a home winemaker and insisted that he made an excellent wine. To convince us he went home and returned with one of his wines. Pouring a glass, one of our workers smelled it and immediately recognized the "Wine" as oxidized and it even tasted like vinegar. Here was an individual who made "Wine" but never recognized it's quality. His wine experience was with his own wine. Such experiences give some credence to the old saying, "Winemakers learn to live with their own wine— good, bad, or indifferent."

Home winemakers have the luxury to experiment with various amounts of wine, from lots as small as one gallon to 50 gallon quantities. Sometimes the amount of wine will depend on the availability of fruit. Recently we produced one gallon of white wine from our arbor patio grapes, varieties unknown, and entered it in a winery annual fund raising home winemaking competition. The wine proceeded to win a first place. This wine probably could not be duplicated again.

Amateurism is diminished and professionalism enhanced whenever a home winemaker uses superior grapes, employs appropriate apparatus and chemicals, and develops accurate, consistant written cellar notes.

Some home winemakers belong to organized wine clubs, but they probably represent a small fraction of all home winemakers. Club members can benefit by sharing their experiences. Portions of this article represents generalizations from the activities of one wine club, but there are similar groups in Los Angeles, Orange, Contra Costa, El Dorado and Napa Counties. Because national or state wide winemaking associations do not exist, inquires regarding local clubs may best be found by consulting the yellow pages for the name of a local wine fermentation or supplier shop. Some of these groups are guided by fermentation supply shops and others are independently managed.

Purpose of Wine Clubs

The Sacramento Home Winemakers Club was founded in 1973 and Incorporated in 1974 as a nonprofit organization. The general purpose of the nonprofit club was to promote interest in the art of winemaking by the amateur. Specific purposes were to organize, conduct and/or attend discussions, lectures, field trips, experiments, and competitions (1).

Critical Analysis of Wines. SHWI's first President was George Leone. The meetings were held in a neighborhood public school where tasting of wine was not permitted so the meetings were moved to members' homes. Sensory evaluation of the members' wines was an important part of the clubs activities. In 1976, a more formal meeting place was selected. This facility, the VFW Hall in West Sacramento, made it possible to evaluate wines at monthly meetings. This practice was started during Leone's tenure of office and continues to this day.

Joe Kramer, the second club president (1974), instituted a formal evaluation of wines by inviting a noted wine authority, Darrell Corti, to comment on members' new wines. Initially Corti evaluated all the wines (red, white, fruit, exotic and sparkling wines) produced by the club's members. This was done once a year. As the membership and wines increased, two sessions were set aside in the spring of each year, one for red wines, and the other for whites. Other wine experts were invited to evaluate the fruit and exotic wines. The wine analysis meetings of members' new wines continues to this day. These meetings attract the largest attendance.

Obviously if SHWI members observe, listen, and act on the experiences gathered in these tasting and wine evaluation sessions, their wines will improve. At the February 19, 1992 club meeting, Darrell Corti evaluated 18 Zinfandels, four Cabernet Sauvignons, three Barberas, two Charbonos, two Petite Sirahs, and one each of Carrignane, Carmine, and Concord Wines. In addition to the usual critical and pointed remarks, Corti also made two general observations regarding the problems of a few wines. The first was a caution about using Montrachet yeast. This yeast may produce a stinky wine, if the grapes had been treated with sulphur in the vineyard. A suggestion offered to correct the problem was to use copper tubing, aeration, and racking proceedures to remove the sulfur by-products. The second observation was a caveat regarding the use of oak chips. Overly oaked wines may produce a "mushroom-like" odor. Such an odor detracts from the floral and fruity aroma of young wines.

Furthermore, oak chips tend to absorb color and tannin. Corti pointed out that the primary function of oak barrels was to provide a controlled oxidation, rather than imparting an aroma or flavor in the wine (2).

Concurrent with wine evaluations, Ralph Stellrecht, the third club president (1975), invited guest speakers from the University of California, Davis. One presentation introduced club members to the Davis 20 Point Wine Evaluation Form. This tool was ultimately adopted as the formal way to judge members' wines.

Homemade WIne Competitions. It was Joe Kramer's idea to include a picnic with socializing, wine sampling, and wine judging. Jack Namle the newsletter editor provided the name, the "May Jubilee," for this social activity. The first May Jubilee wine judging which took place in 1974 was informal. During Ralph Stellrecht's presidency (1975), the judging was still informal, with club members serving as judges, and the "Davis 20 Point Scale," was used to evaluate the wines. This first attempt at competitive evaluation of a wine against a specific standard, although interesting and informative, proved not to be universally accepted by the membership. Changes were made in 1976 and again in 1977 so that in 1978, under the direction of Gerald Cresci, a more formal May Jubilee judging was conducted with non club members as judges. The May Jubilee judging provided a training ground for members to test their wines, to develop judging organizational skills, and to supply a cadre of personnel who would provide leadership for other competitive judgings. It was during Ed Dulce's presidency (1976) that SHWI became involved with the Sacramento Wine Festival which included an exhibit booth during the "Festival." In 1978 and again in 1979, under the guidance of Gerald Cresci, the SHWI conducted a homemade wine judging for the Sacramento Wine Festival. The Festival was discontinued when the distributor, Vintage Wine House, gave up the sponsorship.

During Ralph Barnett's presidency (1977), Jack Namle made arrangements to co-sponsor with the California State Fair, a state-wide homemade wine competition. The competition was professionally organized, administered, and successfully conducted. Every detail for the proper administration of the judging was undertaken. Namle consulted with George Cook of the U. C., Davis Cooperative Service. Cook was familiar with details of the successful California State Fair Commercial Wine Judging that was initiated in 1855 and continued until 1967. The trade publications credit it as the premier competition in America during its tenure. In 1985, after a 17 year absence, the competition returned and ranks among the top judgings of American Wine competitions.

Prominent judges were sought for the first California State Fair Homewine competition. These included: Leon Adams (Author), Peter Brehm (Wine and The People), Stanley D. Burton, M.D., Albert D. Webb (U.C. Davis), Roger Bolton (U.C. Davis), Charles Myers (Harbor Winery), Darrell Corti (Sacramento Wine Merchant), Tom Farrell (Franciscan Winery), Bernard Rhodes (Berkeley Food and Wine Society), Bell Rhodes (Berkeley Food and WIne Society), Robert Adamson, M.D., and Dorothy Adamson. Score sheets for judges were developed to accomplish the following: first, should the wine be retained or eliminated; second, if retained, what award should it receive; and third, the judges were requested to write helpful comments regarding each wine. Three judges were assigned to each panel. The panels used the rule of two; i.e. if two judges scored a wine the same, it would receive that award. The organized structure of the competition has remained the same from 1977 to the present. Refinements have been made; the most notable is that all paper work is now processed by computer. Home winemakers are most appreciative of the returned evaluation sheets with the judges' comments. Nearly all of the larger competitions employ this technique. The judges' comments are most helpful to home winemakers in their quest to improve their wines.

Gerald Cresci, in 1987, negotiated the homemade wine competition at the Lodi Spring Wine Show. This competition is unique on several counts; One, judging is held during the "Wine Show" and the attendees can observe the judges in action. Two, the "Wine Show" is a two day event with the white wines being evaluated the first day and the red wines are evaluated the second day. Third, the wine competition entry fee is lower than the major homewine competitions and is held prior to the California State Fair, Sonoma Harvest Fair and the Orange County Fair homewine competetions. Keeping fees low is done deliberately to permit the winemakers to field test their wines without a significant financial outlay which is important to home winemakers because of their limited production and the cost related to entry fees and shipping. The advantages of entering in county or local competition are (1) the entry fees tend to be nominal and (2) there are minimal shipping costs.

Club Lectures and Seminars. SHWI has educated their members by inviting informative speakers to their monthly meetings, conducting wine seminars, and holding round table discussions.

Over the years Dick Fish, Professor of Chemistry at California State University, Sacramento, has instructed the club on such topics as acidity of wines and acid adjustments; wine stabilization and clarification of wines with fining agents; component analysis and the aroma wheel.

Representatives from Robert Mondavi Winery have conducted sessions on wine component identification, effect of soil composition on varietal characteristics of Sauvignon Blanc wines, and oak wine sample recognition.

Other informative sessions include seminars by David Storm who conducted a session on winery sanitation and Scott Harvey who discussed "Ice Wine." Other speakers have come from the wine industry with discussions on winemaking equipment and the use and repair of barrels. Club members have conducted sessions on varietal identification. Members' varietal wines have been paired with commercial varietal wines to aid in wine improvement.

Saturday seminars have been conducted on red wine, white wine, and champagne production. Lisa C. Van de Water conducted a wine analysis seminar on such topics as fining agents, yeast, sulfur dioxide, wine spoilage microbes, and propagation of malolactic bacteria, and concluded with an analysis of members' problem wines.

On another occasion Bruce Rector presented a seminar which contained such topics as:

*Consideration of the individual steps of the crush and the stylistic options.
*Tools to talk about aroma and taste sensations of wine.

*A new way of conceptualizing flavor manipulation in the vineyard.
*How to live with the whims of nature.
*How to live with the whims of technology.
*Fining practices-theory and application.
*Filtration
*Spoilage
*Understanding sulfur dioxide to avoid spoilage.
*Stability in the bottle (3).

Not to be overlooked are the round table discussions by the members on problem wines and the consultations with experts to resolve the problems.

Field Trips. The least often activity undertaken, but nevertheless conducted, are the field trips. These include trips to barrel builders, wineries, and vineyards. Some of the trips include commercial wine sampling and lunches or dinners.

Social Activities: May Jubilee/ Octoberfest/ Christmas Parties. At such events as the May Jubilee, members picnic and sample wines. There is the anticipation of the formal wine judging results. Everyone is there to have a good time. Octoberfests are dinner events with winery presenters. Commercial wines are matched with food and the winery representative discusses the pairing. Of course, homemade wines are also available. Usually the Christmas parties are more formal events in which retiring club officers are recognized and new officers installed. WIne is invariably sampled. Sometimes members' wines are brown bagged and votes cast for the three best.

Suppliers Influence

During the early 1970s there were two fermentation supply shops in Sacramento. One known as "The Little Old Winemakers Shop." It remained in operation for several years and then disappeared. The other shop was called "Grandfather's Basement." It had only a brief existence. It was The Little Old Winemakers Shop that indirectly helped to establish the SHWI. In the words of the club's founder and first president, George Leone:

"When it comes to hobbies, I'm a great believer in sharing what I know with others and to learn as much as I can from those who have the same interests. An obvious way to do that is to either join a club or start one, if nonexistant. In early 1973, I visited the winemaking supply store (The Little Old Winemakers Shop) and asked the lady, (Marcha Brown) who ran it, if there was a home winemaker's club in Sacramento. 'No, not that I am aware of,' she replied. 'But if you want to start one, I'll be glad to help.'"

"If I could get the names and telephone numbers of your customers I would be glad to try and start a club," I volunteered, figuring that would be easy enough to do."

"She offered to compile a list for me in a couple of days, and she did. I made fifty phone calls and held an organizational meeting at a school where about twenty people showed up during very bad weather. Routine matters were discussed like aims of the club, dues, officers, etc. A couple of committees were set up, and we were in business. I was elected president at the following meeting and the Sacramento Home Winemaker's Club became a reality" (4).

The first winemaking class I (Gerald Cresci) attended was at The Home Wine Makers Shop located on Fulton Avenue in Sacramento. This class was conducted by Rich Pierson. The text used was Enjoy Home Winemaking-A Guide For the Beginner. Topics in the class included:
* Equipment.
* Winemaking Additives.
* Sugar, Alcohol, The Hydrometer and Sweetness.
* Acid Testing and Wine.
* Working With Your Wine.
* Winemaking Ingredients and Recipes.
* Aging and Bottling Your Wine.
* Making Sparkling Wine.
* Common Problems in Winemaking (5).

Early in the history of the SHWI, the club started a wine supply store. The first manager was John Gabri. In 1975 Elaine and Ralph Housley secured a resale license and the shop was set up in their garage. This team became the wine confidants and advisors to club members. Upon their retirement, in 1981, Ralph Housley and Ralph Barnett teamed up to organize R & R Home Fermentation Supplies (R&R). The club store was closed, and the stock sold to R & R. Eventually, R & R became the only retail fermentation supply store in the Sacramento metropolitan area. The shop's influence on home winemakers has been a dominent force. Like most shops in California they are available on a day to day basis to advise and guide home winemakers with their problems and concerns.

R & R Home Fermentation Supplies acts as a recipient for wine competition entries. This service is a common practice throughout the state. Some fermentation shops actually organize and conduct home wine competitions for county fairs.

Another service preformed by R & R and other shops is that of wine analysis. For a nominal fee winemakers can get tests on wine alcohol, acid, malolactic fermentation, residual sugar, and other component evaluations.

Nationally, the Home Wine and Brew Trade Association conducts an international homewine judging. To guarantee the quality of homewine judges, they administer a wine test to certify judges. This test was developed by Gerald Cresci. After field testing, the test has been administered extensively throughout the United States. Administration of the test is undler the direction of Ralph Housley. For details of the test and certified wine judges should be directed to Ralph Housley 630 Parkstone Drive, Folsom, CA 95630 or to Home WIne and Beer Trade Association, 604 N Miller Road, Valrio, FL 33594.

Suppliers-Home Wine and Brew shops have been a prime mover in the improvement of homemade wines. These shops are permanently established and seem to be much stronger than they where in the early 1970's. As long as the owners' dedication and profitability continues they will remain a strong influence on the improvement of homemade wines.

Wine Education

California adult education in public schools in the 1960's and 1970's offered viticulture and wine appreciation classes. In the fall of 1974 Nellie and I attended a Viticulture/ Wine Study Class conducted by Dan Pratt. The outline of the course included the following:

* Wine Through the Ages.
* Grape Varieties and Vines
* Commercial Wine Production-Steps in the Conversion of Grapes into Red and White wines.
* Commercial Wine Production-Special Processes (Sparkling, Fortified, Pop Wines).
* Home Wine Equipment
* Home Wine Production-Fermentation to Storage.
* Selecting Dinner WInes.
* Selecting Specialized Wines-Dessert/Sparkling.
* Selecting Imported Wines.
* Wine in the Home-Collection, Cooking.
* Judging and Tasting Wines.
* Field Trips (6).

Rarely does one find viticulture wine courses in California public adult education schools today.

Most of these types of courses are now offered in the California Community Colleges. In the Sacramento-San Joaquin Valley areas the viticulture wine courses are offered in the community college Agriculture or Food and Beverage Programs depending upon the educational emphasis. Such courses are offered at Sierra College, Consumnes River College and San Joaquin Delta College.

California State University, Sacramento offers a nine-week, non-credit course through the Extended Learning Program called "The Appreciation of Wine and Its Chemistry." This course has been taught for the past fifteen years during the Fall semester by three Chemistry faculty members. The course includes all aspects of winemaking with students taking part in picking, crushing, pressing, fermentation, chemical analysis, fining, filtering and bottling. Sensory evaluation of commercial wines are performed during each class with special emphasis on varietal characteristcs, component analysis and aroma recognition. The course also includes a guest speaker, usually a commercial winemaker; a field trip to one of the viticulture regions in northern California and a spring picnic where current and previous class winers are sampled (7).

At California State University, Fresno, Dr. Barry H. Gump teaches a course, which should be especially useful for home winemakers, entitled "Analytical Techniques for Grape, Must and Wine Analysis." Objectives of the course include: (1) analytical methods to provide essential information for the characterization of grapes being brought into the winery and (2) measurements useful for following the process of fermentation, storage, blending and bottling of winery products (8).

Extension Programs at U.C. Davis offer courses that have attracted home winemakers. These include:

> * Legal Aspects of Establishing a Winery. This course covers the legal requirements and agencies involved in the bonding/approval process.
> * Introduction to Sensory Evaluation of Wine.
> * Managing the Home Vineyard. This course covers ground cover, fertilizing, frost, canopy management, irrigation, pest and disease con trol, harvesting, and pruning.
> * Successful Home Winemaking.
> * Successful Small Scale Winemaking. This course is offered for advanced home winemakers who produce small lots wine using minimal equipment.
> * Introduction to Wine Analysis. This course is designed primarily for home winemakers. It includes these crucial tests: free and total sulfur dioxide, volatile and titratable acidity, pH, malolactic paper chromatog raphy, sugar, and present alcohol (9).

The University Extension attracts some home winemakers, but the limiting attendance factors are their location and the per course fees.

Grape Source

The SHWI in the summer newsletter publishes a grape source list. This list includes the name, address and telephone number of the vineyards that have grapes available for home winemakers. Also listed are the types of grapes, their cost and type of equipment available to process grapes.

This practice was initiated by Marsha Brown who supplied fresh grape sources for her customers. Many of the names that appeared on her list and in many of the SHWI newsletter are the same. Club members routinely purchase grapes from these growers, both as individuals and in groups.

It should be noted that the advantage of group grape purchases, is the sharing of equipment, experiences, and savings.

Some supplier shops locate grapes and attempt to match their grapes with clients' needs. However, these grapes are more expensive because of the number of middlemen involved in the distribution process. This is a service to home winemakers who do not have the time or inclination to seek out their own source of grapes.

Many home winemakers have their own backyard vineyards which produce sufficient quantities to meet their needs.

Summary

1. I have attemped to answer the question as to whether home winemaking has improved in the last few years, by using an experiental technique. There isn't a shred of scientific evidence to demonstrate a positive response. Observations and the

experience of the author are the basis of the comments presented. If one disagrees with the conclusions, then it is possible the observations were faulty or misinterpreted.

2. Some home winemakers have a natural talent for producing good wines consistently. Other home winemakers will violate all good production techniques and develop acceptable wines, but this is a rare occurrence. More frequently the practice is to follow a recipe. A recipe procedure gives the winemaker a feeling of security.

3. Home winemakers can experiment with the wines they produce in terms of quantity and the personal preference type of wines. To do this good practices must be followed and the ability to use winemaking knowledge is important. Knowledge of good winemaking practice such as, superior grapes and fruit varieties, appropriate equipment, good cellar practice and equipment, etc. will result in the production of high quality wines.

4. Only a small number of home winemakers belong to organized winemaking groups. These groups are probably the single most influential element in the improvement of home produced wines.

5. Home winemaking clubs have been established in a number of California counties which include Sacramento, Los Angeles, Orange, Contra Costa, El Dorado, and Napa. Some of these clubs are guided by fermentation or home wine supply shops and others are independently managed.

6. The Sacramento Home Winemakers Club is a nonprofit incorporated organization which has been one of the leaders in the improvement of homemade wines. General and specific objectives of these clubs are to (a) promote interest in the art of winemaking by the amateur, and (b) organize, conduct and/or attend discussions, lectures, field trips, demonstrations, and competitions.

7. Evaluation of club members' wines is undertaken on a regular basis. Critical analysis of wines is performed by wine experts and club members. This arrangement has led directly to the improvement of members' wines.

8. Homemade wine competitions are developed on a national, state, regional, and local basis. The written evaluations provide home winemakers with immediate feedback on their wines. These competitions are the single most influential factor which improve homemade wines.

9. A number of different techniques are employed by clubs to formally educate members regarding wine, these include: guest speakers at meetings, wine seminars, and round table discussions.

10. Less formal education methods are employed to instruct club members such as field trips and special social activities such as May Jubilees, Octoberfests, and Christmas parties.

11. Winemaking information acquired via club "networking" serves to improve members' wines.

12. Second only to the influence of clubs and of their activities on the improvement of home wines is the role of winemaking supply shops. Suppliers or fermentation shops become on a day to day basis the confidants and advisors to home winemakers. These shops in addition to furnishing supplies, equipment, and materials

to home winemakers provide an immediate source of information to solve their problems. As long as these establishments are profitable, they will continue in this valuable role.

13. Wine education in public adult education programs, community colleges, The California State University and the U. C. Extension programs is available to home winemakers. The California Community Colleges have taken over the wine education offerings of public adult education schools. It is a rare occurance when an adult education program will offer a wine education class. Community colleges offer them in their agricultural or food and beverage programs. The California State University has a limited number of offerings through their Extended Learning Programs. The most extensive offering may be found in the U.C. Extension program. Unfortunately, although wine education courses are available, they are not an important element in the improvement of home winemaking at this time.

14. The availability of a grape source is an aid and a convenience to home winemakers in their wine production. The quality of fruit is critical for improving wines.

15. Homemade wines have improved. Given all the limitations of home wine production, it is remarkable that good wines are made. Limitations notwithstanding a majority of home winemakers continue to produce increasingly acceptable wines.

16. It is a fact that home winemaking is improving—this is not a fantasy.

Literature Cited

1. Constitution and Bylaws of Sacramento Home Winemakers, Inc. amended November 18, 1987.
2. The Grape Vine, Sacramento Home Winemaking, Inc., March 26, 1992.
3. Rector,Bruce.; Napa Valley School of Cellaring, Sacramento Home Winemakers, July 13, 1985.
4. Leone,George.; Unpublished Autobiography, June 1987, p. 355-356.
5. Frishman,Robert.and Frishman,Eileen.; Enjoy Home Winemaking-A Guide for the Beginner, The Winemaking Shop, 1972.
6. Pratt,Dan.;Viticulture (Wine Study) 401 John F. Kennedy Adult Education Center Sacramento City Unified School District Fall Quarter 1974.
7. Fish,Richard.; California State University, Sacramento, The Appreciation of Wine and its Chemistry, notes submitted to the author June 1992.
8. California State Universtiy, Fresno, Course Syllabus, Enology 114 (Analytical Techniques for Grape, Must and Wine Analysis) Circa 1991.
9. Practical Winery & Vineyard, Calendar, January - February 1991, p. 50; May - June 1991, p. 69; January - February 1992 p. 44.

RECEIVED May 6, 1993

INDEXES

Author Index

Affiliation Index

Subject Index

Production: Meg Marshall
Indexing: Deborah H. Steiner
Acquisition: Anne Wilson
Cover design: Alan Kahan

Printed and bound by Maple Press, York, PA